PROCESS CHROMATOGRAPHY
A Practical Guide

CONTRIBUTORS

Adner, N.
Berglöf, J.
Hagel, L.
Hedman, P.
Markeland-Johansson, C.
Johansson, H.
Jones, B.
Larsson, B.
Lieber, A.-K.
Nyström, L.-E.
Persson, O.
Petterson, T.
Sofer, G.
Sparrman, M.
Wahlström, H.
Whitney, D.*
Östling, M.

*All contributors from Pharmacia, Uppsala, Sweden, except Whitney, D. and Sofer, G., from Pharmacia LKB Biotechnology, Piscataway, New Jersey, USA, and Jones, B., from Pharmacia LKB Biotechnology, UK.

ACKNOWLEDGEMENTS

The editors wish to thank Dr Rainer Eketorp of KabiVitrum AB and Dr Stuart Builder of Genentech, Inc. for reading this book and making suggestions for its improvement.

PROCESS CHROMATOGRAPHY

A Practical Guide

Edited by

G. K. SOFER
Pharmacia LKB Biotechnology Inc.
Piscataway
New Jersey
USA

and

L.-E. NYSTRÖM
Pharmacia LKB Biotechnology
Uppsala
Sweden

ACADEMIC PRESS
Harcourt Brace & Company, Publishers
London San Diego New York
Boston Sydney Tokyo Toronto

ACADEMIC PRESS LIMITED
24-28 Oval Road
London NW1 7DX

United States Edition published by
ACADEMIC PRESS INC.
San Diego, CA 92101

Second Printing 1994

British Library Cataloguing in Publication Data

Process chromatography
1. Biochemistry. Chromatography
I. Sofer, G K II. Nyström, L-F
574.19'285

ISBN 0-12-654268-6

Phototypeset by Dobbie Typesetting Limited
Printed in Great Britain by Antony Rowe Ltd, Chippenham, Wiltshire

Contents

Foreword

Protein purification plays a fundamental role in modern biotechnology. It is used in research, development and production. Many workers in the field received their introduction to chromatography by reading *Gel Filtration in Theory and Practice*, followed by other monographs on adsorptive techniques such as ion exchange and affinity chromatography. The purpose of this book is to go beyond the description of individual chromatographic methods to teach the integration of steps, both up- and downstream, leading to a complete process for the purification of a desired protein.

Since the invention of partition chromatography nearly a half century ago, which earned Archer Martin and Richard Synge the Nobel Prize in Chemistry (1952), this separation technique has been applied to an enormous number of molecules and feedstocks. The introduction in 1959 of soft gels or resins which could be used in packed beds dramatically increased the capacity and scalability of the technique, with the result that liquid chromatography has become a preparative tool as well as an analytical one (Porath and Flodin, *Nature* **183** (1959) 1657). The scale of operation now stretches over nearly nine orders of magnitude with respect to the amount of material prepared and at least six orders of magnitude with respect to the volume of the packed bed. The first, and still most common, chromatographic separation modes were ion exchange and gel filtration (also called size exclusion). They are effective alone, but even more so in combination, for purifying small to moderate-sized molecules. For larger more complex biopolymers, additional methods such as the recently popularized affinity chromatography are also frequently used to great advantage.

Most biochemistry laboratories today could not function without at least one type of chromatography. A pharmacologist may use it to determine the concentration of a drug in the circulation. A toxicologist may use it to distinguish a safe chemical from a toxic analogue. A molecular biologist may use it to separate nucleotides from polynucleotides in the preparation of a probe. An enzymologist may use it to purify a protein in order to study its structure and function. It may be used to prepare an industrial enzyme and will certainly be used in the purification of a pharmaceutical protein. The applications are myriad but the principle is always the same: use of some molecular property of the desired product and its contaminants/impurities that interact to a different degree with a stationary phase to achieve a differential migration. The molecular property may be size, charge, hydrophobicity or some very specific interaction as in affinity chromatography. The choice depends upon the starting material and the goals.

Chromatography has become such an integral part of the biotechnology industry that we do not know of a single biopolymer product that is purified without it. Most regulatory authorities expect that at least one chromatographic step be used in the production of a protein-based pharmaceutical. This is primarily because it is impossible to achieve the currently expected levels of purity using conventional nonchromatographic separation methods such as fractional precipitation.

In the continuing evolution of chromatography, several trends stand out. Chromatography media are being developed which have increased stability. This includes both chemical stability to increase useful life and physical stability to increase maximum operational flow rates and improve *in situ* regeneration.

There is an increased emphasis on sanitation. More resins are being made base stable since strong alkali is widely accepted as a safe and effective means of depyrogenation, sanitization and cleaning. At the same time, non-basic cleaning, sanitizing and storage solutions are being developed for use with separation media that cannot withstand high pH, such as immobilized monoclonal antibodies. Suppliers are designing systems (equipment and media) capable of sterile operation.

Until recently, the use of the high resolution achieved with small particle size (e.g. 10 μ) resins was dominated by analytical separations. With advances in polymer chemistry and the technology of resin manufacture, a greater variety of preparative media with smaller and more uniform average particle size are being developed. Thus, more investigators are able to take advantage of the superior separating power and speed of operation which can result.

Columns are being designed for ever larger operations. It is no longer unusual to see columns with diameters over 1 m and working volumes over 1000 l. New designs ensure that the structural and flow distribution demands of such large columns are being met. Very large columns will be required for the purification of high volume products such as bovine growth hormone with a potential market need in the USA estimated at 100 000 kg year^{-1} and albumin with an even larger worldwide current use (about 250 000 kg year^{-1}).

An area of significant new development is chromatographic monitoring and control. Many new sensors along with computer workstations and the associated control software are coming on stream. At the analytical scale, such systems are being used to aid method development and optimization of separation. At the production scale, they are used for unattended operation, precise control and reproducibility, and documentation for GMP operations.

The development of recombinant DNA and hybridoma techniques, 'the new biotechnology,' presents a great opportunity in the integration of upstream and downstream operations. It is receiving a great deal of attention with much accompanying success. It is fair to say that if the downstream processing scientist waits for the whole broth or conditioned media to 'arrive on his doorstep' to begin his purification, he has missed some of his greatest opportunities. For example, once a molecular biologist sequences a gene as part of cloning it, much information about the target molecule becomes available. This includes the size, sequence, composition, pI and potential glycosylation sites, all of which can be used to begin developing purification strategies even before the protein may have been expressed. In addition, there are a number of notable examples in which specific nucleotide sequences have been incorporated into the expression vector to significantly ease downstream processing. Collaboration with the fermentation scientist may include determining how media components and host proteins behave chromatographically and their effect on the product itself. The rapid development of the process for plasminogen activator (rtPA) relied to a substantial measure on such cooperation.

Pharmaceutical proteins are being made at unprecedented purity levels which sometimes exceed 99.999% (or impurity levels less than 10 ppm). It is now sometimes more difficult to measure and prove these levels of purity than it is to achieve them. The chromatographic challenge of today is to reach these levels of purity with fewer more specific separations at higher yield and economy.

We expect that many of these achievements are just 'around the corner.' In addition, completely new uses of chromatography are most certainly being worked on. If this field continues to develop at its current pace, the examples in this book will be outdated soon. The principles taught, however, should last many years.

John M. Curling
Uppsala
Sweden

Stuart E. Builder
San Francisco
USA

1 Introduction

This handbook is a guide and reference source for the development of an economical chromatographic purification process for biological molecules, especially peptides and proteins. Taking into account the multiple factors that are interdependent (e.g. the source of raw material, isolation steps, purification, scale up, product analysis and regulatory considerations), we present a systematic approach for designing a purification process, along with guidelines for eliminating common pitfalls. Appendices include background information on chromatographic techniques, product analysis, regulatory considerations and column packing.

Chromatography is generally recognized as the technique which allows the highest degree of purification of biomolecules. The US Bureau of Biologics, for example, recommends that at least one chromatographic step be used in the purification of monoclonal antibodies.[1] In contrast to techniques such as precipitation, electrophoresis, isoelectric focusing and ultrafiltration, chromatography does not involve heat generation or major shear forces. Conditions close to physiological can be used to maintain biological activities during chromatographic steps. Process-scale chromatography is used today in the purification of biologicals from natural sources (e.g. albumin from plasma) and biologicals produced by modern biotechnology (e.g. monoclonal antibodies and products of recombinant DNA technology such as interferons, vaccines, insulin, growth hormones and tissue plasminogen activator).

Many production-scale separation processes for commodity chemicals have been described in detail. Equilibria are known, and mass transfer rates can be calculated from correlations discussed in many textbooks. For biological macromolecules, however, far less is known about chemical equilibria and mass transfer rates, so it is not possible to calculate the outcome of a tentative separation protocol from existing data. A basic understanding of the fundamentals of chromatography, however, is useful to interpret

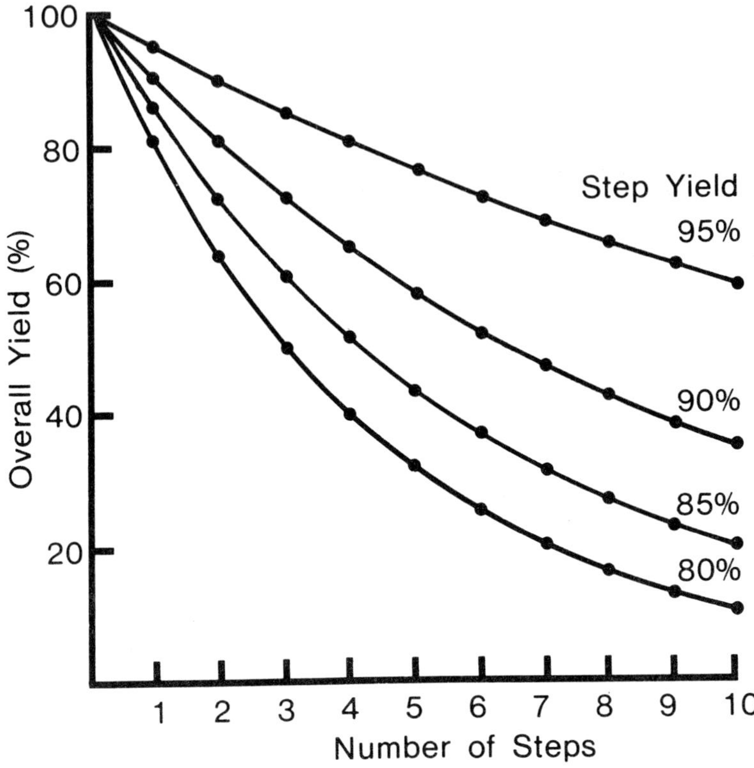

Figure 1 Loss of yield compared with number of steps.

Table 1. Chromatographic criteria.

Criteria	Analytical	Process
Sample	Minimize	Maximize
Production	Cycles h^{-1}	Kg h^{-1}
Resolution	Number of peaks	Recovery

Table 2. Major factors in the large scale chromatography of proteins.

Resolution (selectivity, efficiency)
Recovery
Throughput
Reproducibility
Stability (life length of adsorbent)
Hygiene (maintenance)
Convenience
Economy

laboratory experiments and to predict which modifications may improve a separation.

The optimal use of chromatography is important when purifying products in a cost-effective manner. The high degree of purity required for therapeutics necessitates the use of more than one chromatographic step, but careful optimization of the purification scheme will minimize the number of steps necessary to reach required purity. *Figure 1* shows how the yield decreases as the number of purification steps increases. If, for example, each individual step has a 90% yield, the overall yield after four steps is less than 70%. (In reality, the steps in a process seldom have the same yield. A typical example might be four steps with yields of 85, 90, 70 and 90%, which would give an overall yield of approximately 50%.) A cost-effective process requires that each step be performed in such a way as to ensure maximum throughput, maximum yield and maximum recovery of biological activity.

The factors governing industrial and analytical chromatographic processes are different (see *Table 1*). In the case of industrial chromatography, selectivity and efficiency (and, as a result, resolution) are optimized for throughput and recovery calculated by quantity of biologically active product purified per hour to a previously set specification.[2] *Table 2* shows the factors that are of major importance in the large scale chromatography of proteins.

The establishment of a purification protocol intended for production requires several steps. First, the initial feed is characterized and prepared for chromatography. Each chromatographic step is optimized on a laboratory scale using chromatographic supports that are readily scaled up and capable of multiple cycles. The steps are then put together in a cost-effective protocol. During scale up, appropriate equipment is employed and established guidelines are followed. Finally, hygiene protocols are developed to ensure product safety, long gel life, and hence cost-effective chromatography.

REFERENCES

1. US FDA. *Points to Consider in the Manufacture of Monoclonal Antibody Products for Human Use — 1987.* Office of Biologics Research and Review, Center for Drugs and Biologics (1 June 1987).
2. Janson, J.-C. and Hedman, P. On the optimization of process chromatography of proteins. *Biotechnology Progress* **3** (1987) 9-13.

2 *The Choice of Host*

The choice of host for the production of a biological molecule such as a protein will have a significant effect not only on the isolation steps, but also on the purification of the product.[1] Two considerations are overriding in this choice: the preservation of the product, and its isolation and purification to the required degree of purity. Today, products of potential commercial value are still being produced from more traditional sources such as plasma and tissue. But they are also being produced by modern recombinant DNA technologies in such hosts as bacteria, yeast, filamentous fungi, and insect and mammalian cells.

For a recombinant expression system, isolation, as well as the choice of the initial chromatographic method, will be affected by the location of the product. If the product is produced in bacteria by recombinant DNA methods, it may be secreted into the fermentation broth, directed to the periplasma space,[2] or found in insoluble aggregates (inclusion bodies) within the cytoplasm (see *Figure 2*).

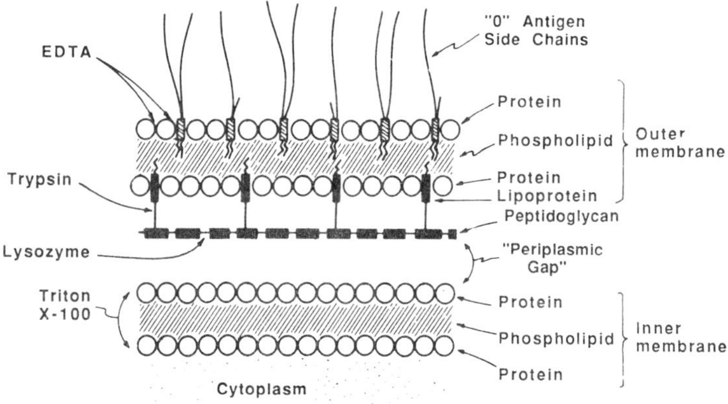

Figure 2. Schematic drawing of a bacterial cell membrane indicating the substances present in the cell lysate. Also indicated are potential releasing agents.

Kane and Hartley[3] have suggested that inclusion bodies, refractile structures that contain proteins and nucleic acids but no membranes, develop as proteins aggregate around the protein translation system. Others have hypothesized that the inability of *Escherichia coli* to glycosylate proteins causes aggregation by exposing normally shielded hydrophobic portions of proteins. The isolation of proteins from inclusion bodies often leads to problems with refolding of the protein, correct reformation of disulfide bonds, and thus recovery of activity. It may be advantageous to avoid the formation of inclusion bodies by cloning the gene into a secretion vector. Alternatively, if the protein has a proline as the second or third amino acid, it may be obtained in a soluble form through use of the method of Dalboge *et al.*[4] Through use of this system the overproduction of a precursor of human growth hormone in a soluble form was achieved in *E. coli*. A synthetic N-terminal presequence is tailored to enable the specific removal *in vitro* by the enzyme dipeptidylamino peptidase I (DAP I). The amino terminal extension also allows the use of very specific purification steps. Schein *et al.*[5] found that lower growth temperatures favor the production of soluble recombinant proteins in *E. coli*.

Another example of the avoidance of inclusion body formation comes from the work of Moks *et al.*[6] They have described the production of biologically active human insulin-like growth factor I (IGF-I) on a 1000 l scale from *E. coli*. IGF-I is produced as a fusion product containing the IgG-binding domain derived from protein A. The fusion product is secreted into the growth medium and subsequently purified on IgG Sepharose® Fast Flow, after which the fusion product is cleaved specifically at the peptide bond between the IgG-binding and the IGF-I domains.

Takahara *et al.*[7] have shown that active human superoxide dismutase can be secreted across the inner membrane of *E. coli* into the periplasmic space.

The problems involved in recovery of biologically-active heterologous peptides or proteins from inclusion bodies in *E. coli* are clearly surmountable. For example, Genentech, Inc., has filed a patent for the "purification and activity assurance of precipitated heterologous proteins."[8] Immunex has used acid extraction of interleukin-1 from inclusion bodies in a way that causes the precipitation of most of the *E. coli* proteins.[9] There are some advantages to having the product in inclusion bodies. An overproduced protein in inclusion bodies is in an insoluble form that may delay proteolytic degradation.[10] In addition, the protein is more concentrated than if it was secreted into the fermentation broth. Unigene[11] has shown that peptide hormones representing greater than 10% of total cellular protein can be produced in *E. coli* as a fusion protein complex. The complex is isolated in insoluble inclusion bodies following *E. coli* cell lysis. The fusion protein is purified prior to proteolytic liberation of the peptide hormone, and the carboxyl terminal amino acid is amidated to produce the biologically active hormone. (It is noteworthy that heterologous proteins have been produced in *E. coli* forming up to 50% of total protein.)

Recent developments in understanding the yeast secretion pathway have enabled biotechnologists to take advantage of the desirable features that yeast offers as host for the production of heterologous proteins: yeast lacks detectable endotoxins; it is generally recognized as a safe organism; it is not a pathogen for humans; large-scale production and down-stream processing of yeast are well established. One group found that a culture could be scaled up from shaker flasks to a 240 l vessel in approximately 100 days. Compared to mammalian cell culture, yeast fermentation is much cheaper, since yeast cells grow rapidly in a relatively simple culture medium.[12] A list of proteins and peptides produced in yeast using α-factor for secretion of heterologous proteins is given in Ref. 11. Yeast secretion systems also facilitate disulfide bond formation and glycosylation. However, glycosylation in yeast is not identical to glycosylation in mammalian cells, and it may not always be possible to produce heterologous glycoproteins with proper biological activity in yeast.

The yeast *Saccharomyces cerevisiae* normally secretes only 0.5% of its own proteins into the growth medium, and most of these secreted proteins are larger than 50 000 daltons.[13] The fact that only a few proteins are secreted simplifies the purification of a secreted heterologous protein. Hepatitis B virus core antigen (HBcAg) has been expressed in the yeast *S. cerevisiae* as approximately 40% of the soluble yeast protein.[14] It is believed that this very high level of expression is due to the lack of any non-yeast-derived flanking sequences that change DNA transcription, translation or mRNA stability. Also the HBcAg polypeptides have a low rate of turnover, possibly due to their aggregation into 28 nm particles that may protect them from proteolytic attack. (For a review on the production of mammalian proteins in *S. cerevisiae*, see Ref. 15.)

Filamentous fungi have also been used for the production of heterologous proteins. *Aspergillus nidulans* has been used to secrete bovine prochymosin.[16] Filamentous fungi have several advantages for secretion of heterologous proteins: There is extensive experience with fermentation of filamentous fungi. They can be grown on simple, inexpensive media in deep batch tanks. They are currently the source of many industrial enzymes. *Aspergillus niger* is GRAS (generally regarded as safe by the FDA). Filamentous fungi have the capability to secrete large quantities of protein into the culture medium; for example, under optimal conditions, *A. niger* secretes glucoamylase, a homologous protein, into the medium to the extent of over 5-20 $g\,l^{-1}$. (However, no known heterologous protein has been expressed at this high level.[17] For further reading, see Refs 18, 19.)

Another potential host for the large scale production of heterologous proteins is cultured insect cells, in which the baculovirus

(BV) is used as a vector system. This system has many of the protein processing mechanisms required for higher eucaryotic proteins, although it is unclear if proteins are glycosylated in the same way as in the normal host. Up to 400 mg of product (β-galactosidase) per liter have been produced with the BV insect cell system. Several interferons and interleukins have also been produced with the BV system. A baculovirus vector was used to express the first recombinant HIV envelope proteins to receive FDA approval for clinical evaluation as a candidate vaccine for AIDS.[20, 21] The volume cost for insect cell culture is comparable to mammalian cell culture, but the production level of product is up to 1000 times greater.[22] Air lift bioreactors from 40 to 100 l have been used to culture insect cells for production of recombinant proteins.[23] For further reading, see Ref. 24.

To obtain a heterologous protein with the same biological activity as the natural protein may even require use of cells from the same species in order to get proper glycosylation, disulfide bridge formation, and other necessary post-translational modifications.

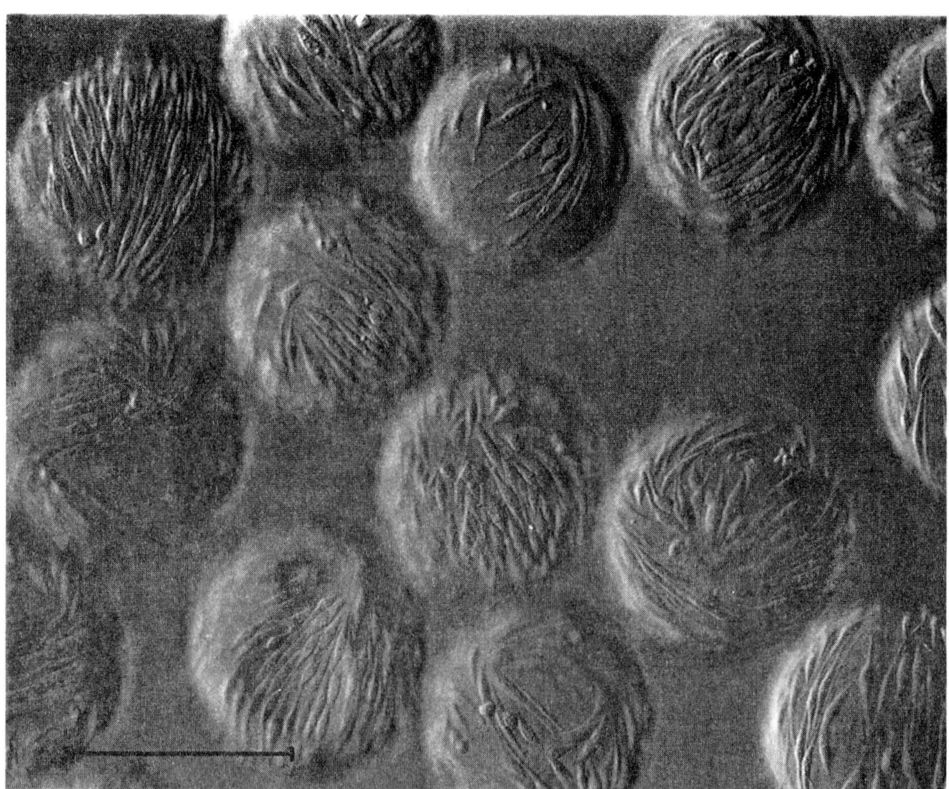

Figure 3. Anchorage-dependent mammalian cells (BHK-21) growing on Cytodex® microcarriers. Scale bar represents 200 μm. (By kind permission of Coopers Animal Health Ltd.)

Successful scale up of mammalian cell culture is thus quite important for the commercialization of biotechnology. Factor VIII has been produced using baby hamster kidney (BHK) cells with the vaccinia virus vector system.[25] Hepatitis surface antigen has been produced from cell culture of recombinant Chinese hamster ovary (CHO) cells.[26]

Some of the problems associated with mammalian cell production systems are cell fragility, slow growth, genetic instability, pathogens and oncogenes, and high media cost.[27] The productivity per mammalian cell is usually quite sufficient, but cell counts are about 100 times less than microorganisms. Increased cell density can be achieved by various methods, including microcarriers. Mammalian expression systems produce product at levels ranging from approximately 1 to 50 $mg\,l^{-1}$.

There are several systems for mammalian cell culture, and some have been scaled up for production of therapeutic proteins. For suspension-adapted cell lines, deep tank systems (either stirred or air lift) and encapsulation techniques have been used. For anchorage-dependent cell lines, the following techniques have been used: microcarriers (see *Figure 3*); roller bottles; immobilized support matrices such as glass beads or ceramics; hollow fibers; encapsulation; parallel flat membranes; static maintenance reactors. For further reading, see Refs 28, 29. Many companies are currently investigating the use of efficient continuous perfusion systems. These systems allow for the continuous removal of product and thus separation from harmful proteases, but defining a 'batch' for regulatory purposes will require serious consideration.

Clearly, there is no one expression system that meets all needs. It appears that the production system should be as close to the natural system as possible. Each production system presents its own set of problems—and benefits. There also appears to be quite a bit of variability from one producer to the next. For a comparison of the biosynthetic capabilities of various expression systems, see Ref. 30.

REFERENCES

1. Pickett, A. and Hardy, K. Comparison of hosts for rDNA. *Applied Molecular Genetics. Proc. 3rd Eur. Cong. Biotechnol.* **4** (1984) 309-313.
2. Glick. B. R. and Whitney, G. K. Factors affecting the expression of foreign proteins in *Escherichia coli. J. Ind. Microbiol.* **1** (1987) 277-288.
3. Kane, J. F. and Hartley, D. L. Refractile formation in recombinant *Escherichia coli. Poster O. 34.* American Society of Microbiology (March, 1987).
4. Dalboge, H., Dahl, H.-H. M., Pedersen, J. and Hansen, J. W. A novel enzymatic method for production of authentic hGH from an *Escherichia coli* produced hGH-prěcursor. *Bio/technol.* **5** (1987) 161-164.
5. Schein, C. H. and Noteborn, M. H. M. Formation of soluble recombinant proteins in *Escherichia coli* is favored by lower growth temperature. *Bio/technol.* **6** (1988) 291-294.

6. Moks, T., Abrahmsen, L., Osterlof, B., Josephson, S. *et al.* Large-scale affinity purification of human insulin-like growth factor I from culture medium of *Escherichia coli. Bio/technol.* **5** (1987) 379-382.
7. Takahara, M., Sagai, H., Inouye, S. and Inouye, M. Secretion of human superoxide dismutase in *Escherichia coli. Bio/technol.* **6** (1988) 195-198.
8. Builder, S. E. and Ogaz, J. R. US Patent Number 4 511 502, April 16, 1985. Wetzel, R. B. US Patent Number 4 599 197 July 8, 1986.
9. Kronheim, S. R., Cantrell, M. A., Deeley, M. C., March, C. J. *et al.* Purification and characterization of human interleukin-1 expressed in *Escherichia coli. Bio/technol.* **4** (1986) 1078-1082.
10. Lowe, P. A. Solubilisation and activation of eukaryotic bodies in *E. coli. Proceedings of Biotech 85*, p. 127.
11. Gilligan, J. P., Warren, T. G., Koehn, J. A., Young, S. D. *et al.* Purification of a fusion protein containing recombinant human calcitonin. *BioChromat.* **2** (1987) 20-27.
12. Yeast beats mammalian cells on manufacturing costs. *Genetic Technology News*, Dec. 1987.
13. Das, R. C. and Schultz, J. L. Secretion of heterologous proteins from *Saccaromyces cerevisiae. Biotechnol. Prog.* **3** (1987) 43-48.
14. Kniskern, P. J., Hagopian, A., Montgomery, D. L., Burke, P. *et al.* Unusually high-level expression of a foreign gene (hepatitis B virus core antigen) in *Saccharomyces cerevisiae. Gene* **46** (1986) 135-41.
15. Kingsman, S. M., Kinsman, A. J. and Mellor, J. The production of mammalian proteins in *Saccharomyces cerevisiae. TIBTECH* **5** (1987) 53-57.
16. Cullen, D., Gray, G. L., Wilson, L. J., Hayenga, K. J. *et al.* Controlled expression and secretion of bovine chymosin in *Aspergillus nidulans. Bio/technol.* **5** (1987) 369-376.
17. Bio/technology Conference "Production Systems: Alternatives for Commercial Production of Recombinant Proteins", New York, May 6, 1987.
18. Cullen, D. and Leong, S. Recent advances in the molecular genetics of industrial filamentous fungi. *TIBTECH* **4** (1986) 285-288.
19. Van Brunt, J. Fungi: the perfect hosts? *Bio/technol.* **4** (1986) 1057-1062.
20. Luckow, V. A. and Summers, M. D. Trends in the development of baculovirus expression vectors. *Bio/technol.* **6** (1988) 47-55.
21. Jasny, B. R. Insect virsues invade biotechnology. *Science* **238** (1987) 1653.
22. Klausner, A. Rearing insect viruses for fun and profit. *Bio/technol.* **3** (1985) 677-680.
23. Weiss, S. A., Belisle, B., DeGiovanni, A. and Godwin, G. The role of insect viruses in biotechnology: past, present and future. *In Vitro* **24** (1988) 53A.
24. Summers, M. D. and Smith, G. E. *A Manual of Methods for Baculovirus Vectors and Insect Cell Culture Procedures. Bulletin No. 1555.* Texas Agricultural Experimental Station.
25. Pavirani, A., Meulien, P., Harrer, H., Schamber, F. *et al.* Choosing a host cell for active recombinant factor VIII production using vaccinia virus. *Bio/technol.* **5** (1987) 389-392.
26. Hershberg, R. US Patent Number 4 624 918, Nov. 25, 1986.
27. Brown, P. C. Considerations in the use of mammalian cells for the expressions of heterologous proteins. *Bio/technology Conference Production Systems: Alternatives for Commercial Production of Recombinant Proteins.* New York, May 6, 1987.
28. Van Brunt, J. Immobilized mammalian cells: the gentle way to productivity. *Bio/technol.* **4** (1986) 505-510.
29. Cartwright, T. Isolation and purification of products from animal cells. *TIBTECH* **5** (1987) 25-30.
30. Bialy, H. Recombinant proteins: virtual authenticity. *Bio/technol.* **5** (1987) 883-890.

3 *Isolation Steps*

Certain key steps must be performed prior to purification by chromatography. Commonly used initial separation steps are shown in *Table 3*. The primary recovery (or isolation) stages are dependent on the source. Tissue is homogenized, and the product extracted. Plasma is filtered. For cell culture systems in which the product is intracellular, cells are harvested, disrupted, and cell debris removed. If the product is secreted, cells are removed, or the secreted product is directly adsorbed from the culture fluid by, for example, an affinity resin. Depending on the feed characteristics, concentration, conditioning or viscosity reduction may be required. In all cases, clarification, i.e. the removal of particulates, is essential to prevent clogging of chromatography columns. There are a number of principles that apply generally to the purification of biological macromolecules, and these must be kept in mind during the isolation steps. Extremes of temperature, ionic strength and pH, as well as organic solvents and foams, should be avoided. Proteolytic activity must be inhibited or controlled.

CELL REMOVAL

The removal of cells can be achieved by a centrifugal separator with 90% recovery of product and 99.99% reduction in cell content.[1] But centrifugation of large volumes can be a technical problem as well as a safety concern because of aerosol dissemination of cells and their products. In addition, it may not always be possible to achieve a 99.99% removal of cells, and filtration will be required prior to chromatography to prevent clogging of the packed media. Cell removal can also be achieved by filtration using filter aids. Various types of filtration systems, such as vacuum filters and pressure filters, can be used. Microfiltration can also be used to remove cells from the feedstream, and is useful to obtain a sterile product free from

Table 3. Steps required prior to chromatography.

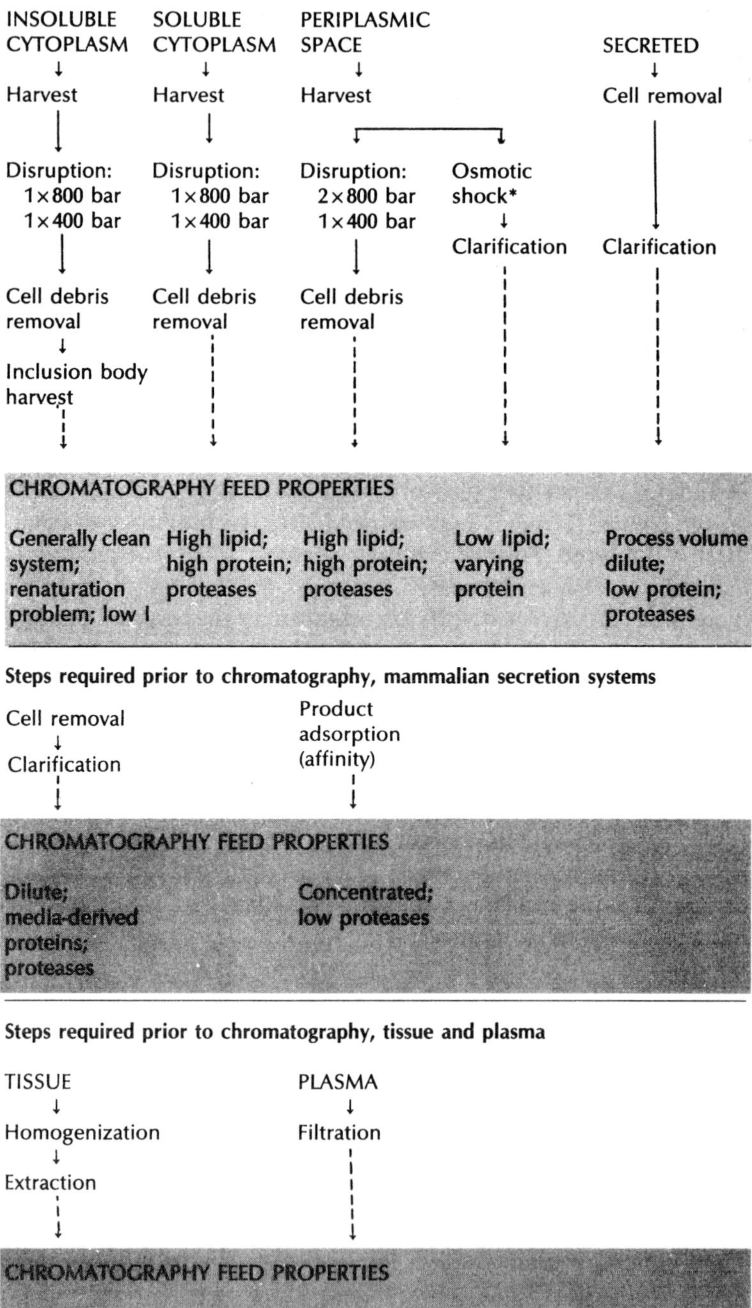

*Osmotic shock will give lower product yield and generally a lower protease level.

cell debris (cross-flow microfiltration, 0.22 μ) A review of cell supernatant clarification is presented by Ball.[2]

CELL DISRUPTION

If the product is intracellular, the proteins must be released by cell disintegration. A review of the techniques available for large scale disruption of microbial cells is presented by Kula *et al.*,[3] and reviews of cell disintegration techniques in general are presented by Scopes,[4] and by Fish and Lilly.[5] A two-enzyme system for selective lysis of yeast has been shown to result in improved product yields and minimized product degradation.[6] Recovery of periplasmic proteins from transformed Gram negative bacteria is described in a patent from Genentech, Inc.[7]

CONCENTRATION

With extracellular proteins, volume reduction and rapid removal of proteases are often critical steps. For example, in the production of monoclonal antibodies by cell culture perfusion systems, the monoclonals are produced in the milligram per liter range. This requires the processing of 100-1000 l of conditioned media, and concentration is a required first step.[8] Process volume reduction may be accomplished by ultrafiltration or by bulk adsorption. Ideally, the concentration step will provide selectivity as well. For example, a bulk ion exchange adsorption step might be capable of concentrating the product and removing proteases at the same time.

The literature describes an evaluation of hollow fiber, plate and frame, and spiral cartridge ultrafiltration systems for reducing process volumes.[7] It is concluded that 'it is important to establish the optimal system, membrane, and process parameters for each product to maximize recoveries and minimize denaturation of the product.' For further reading on ultrafiltration, see Ref. 9.

Concentration is also achieved by precipitation. Precipitation with alcohol, or salts such as ammonium sulfate, has been used traditionally as the initial step of purification schemes. The salts or alcohol are then usually removed by gel filtration or diafiltration. Salts such as ammonium sulfate are quite expensive, and both salts and alcohol present disposal problems. Salt precipitation gives only modest purification and is labor intensive. Recoveries may be less than optimal. To achieve a higher degree of selectivity if salt precipitation is used, it may be possible to salt-out most of the contaminants and then bind the protein of interest in the high salt solution to a hydrophobic interaction gel.

CONDITIONING

Conditioning of the fermentation broth or cell culture fluid may be required to alter some important property of the broth that will allow for increased throughputs during solid/liquid separations (e.g. centrifugation) and reduction of fouling during a whole-broth adsorption step. Broth conditioning techniques include: addition of flocculants; addition of inert filter aids; heat treatment (to reduce viscosity); pH adjustments.[10]

Product degradation is caused mainly by proteolysis. Sometimes it is possible to adjust the pH so that proteolytic enzymes are inactivated and the product stable. If this cannot be done, adsorption of either the product or the proteolytic enzymes should be performed as rapidly as possible, preferably under conditions in which the product is stable, but proteolysis is minimized. However, immobilized enzymes are often more stable than those in solution, so as the feed passes through the column, product could be broken down by immobilized proteolytic enzymes.

VISCOSITY REDUCTION

Viscosity reduction is often a critical initial step. With, for example, *Escherichia coli*, how the cells are lysed will affect viscosity. Sonication yields more fluid to handle; lysozyme disruption will yield a preparation that is too thick to pour.[11] A high nucleic acid content increases the viscosity. The use of an anion exchanger in the presence of a moderately high concentration of salt will, in most cases, allow for the binding of nucleic acids without binding the proteins. Other techniques for the removal of nucleic acids include polyethyleneimine, magnesium chloride or streptomycin sulfate precipitation.

Following product isolation, purification steps can be carried out, but first it is important to characterize the material that is to be put onto chromatography columns.

REFERENCES

1. Lowe, P. A. Solubilisation and activation of eukaryotic bodies in *E. coli*. *Proc. Biotech '85*, p. 127.
2. Ball, G. D. In *Animal Cell Biotechnology* (Spier, R. E. and Friffiths, J. B., eds) Academic Press, New York, 1985, pp. 87–125.
3. Kula, M.-R. and Schutte, H. Purification of proteins and the disruption of microbial cells. *Biotechnol. Prog.* **3** (1987) 31–42.
4. Scopes, R. *Protein Purification: Principles and Practice*. Springer-Verlag, New York, 1982.

5. Fish, N. M. and Lilly, M. D. Review on the interactions between fermentation and protein recovery. *Bio/technol.* **2** (1984) 623-627.
6. Controlled lysis of microbial cells for selective product recovery. *TIBTECH* **5** (1987) 62.
7. Bochner, B. R., Olson, K. C. and Pai, R.-C. US Patent Number 4 680 262, July 14, 1987.
8. Scott, R. W., Duffy, S. A., Moellering, B. J. *et al.* Purification of monoclonal antibodies from large-scale mammalian cell culture perfusion systems. *Biotechnol. Progr.* **3** (1987) 49-56.
9. Cheryan, M. *Ultrafiltration Handbook.* Technomic Publishing, Lancaster, 1986.
10. Rosen, C. G. and Datar, R. Primary separation steps in fermentation processes. *Proceedings of Biotech '83.* Online Publications, Northwood, 1983, pp. 201-224.
11. Personal communication.

4 The Initial Feed

A multiplicity of compounds can usually be found in the initial preparation. Many of these will be closely related to the product of interest; others will interfere with the separation process. One might begin with an analysis of both product and contaminants.

Obviously, focusing on the product is important. Knowledge of the conditions, such as pH, salt concentrations, additives, etc., that preserve the product will be crucial to a cost-effective separation strategy. Some of this information will be obtained during preliminary testing of the starting material.

Some of the characteristics of the starting material that may be useful to determine include: pH, conductivity, concentration of product, total protein and contaminants, viscosity and particulate matter content (*Table 4*). The isoelectric point of the protein and its molecular weight (as well as the isoelectric point and molecular weight ranges of the contaminants) are also important parameters to measure.

An assay for the product of interest will be required to follow the purification. In the production of bovine growth hormone (bGH) by *Escherichia coli* and *Streptomyces lividans*, bGH activity was measured by radioimmunoassay, and acid-precipitable protein was measured by the method of Lowry.[1] A word of caution: the measurement of specific activity based on results from total protein assays (such as Lowry or Biuret) performed in crude preparations

Table 4. Characterizing the initial feed prior to chromatography.

pH
Conductivity
Product concentration
Total protein concentration
Concentration of contaminants (nucleic acids, lipids, pigments, proteases, etc.)
Viscosity
Particulate matter content

may be misleading due to the presence of low molecular weight contaminants that interfere with these assays.

Obtaining a better understanding of the product and the contaminants so that a purification strategy can be selected may be achieved in some cases by electrophoretic titration curves. (For a detailed discussion on electrophoretic titration curves, the reader is referred to the handbook *FPLC Ion Exchange and Chromatofocusing—Principles & Methods* (p. 22) (Pharmacia, Laboratory Separation Division).) In addition, a construction of pH versus net charge curve can be achieved by performing chromatographic titration curves (*ibid.*, p. 25; see also Ref. 2). These two techniques will provide information on the ability of charge-based techniques to separate the product from its contaminants.

The affinity of the product for different group-specific ligands can readily be tested in a batch mode, or small columns can be packed and the product tested for binding under different conditions, such as pH, conductivity, buffer salts, etc.

One company has even established a 'bank' of 10 chromatographic protocols. The 'bank' includes ion exchange, hydrophobic interaction and gel filtration methods with a variety of useful buffer systems. Product stability and separation from major contaminants are examined using high performance columns. Analysis times are short for each step. The media in the 'bank' (e.g. Mono Q®) are those for which corresponding media suitable for scale up (e.g. Q Sepharose® Fast Flow) are available.

A thorough analysis of the product and its contaminants in the initial feed will allow a logical purification scheme to be developed in the shortest time. Consumption of valuable sample will be minimized by performing tests on a small scale and determining what conditions are crucial for maintaining activity of the product.

REFERENCES

1. Gray, G., Seizer, G., Buell, G., Shaw, P. *et al.* Synthesis of bovine growth hormone in *Streptomyces lividans. Gene* **32** (1984) 21-30.
2. Yang, V. C. and Langer, R. pH-dependent binding analysis, a new and rapid method for isoelectric point estimation. *Anal. Biochem.* **147** (1985) 148-155.

5 *Initial Purification Steps*

Of the characteristics described in Chapter 4 (*Table 4*), five are generally considered to have major impact on the initial purification steps: (1) product concentration; (2) total protein and nucleic acid concentrations; (3) conductivity; (4) pH; (5) contaminants.

PRODUCT CONCENTRATION

Depending upon the concentration of the product in the starting material, the first step will be either a concentration step or one intended to give a high degree of purity. Product concentration can vary from below $0.1\,mg\,l^{-1}$ (e.g. in the case of mammalian cell culture supernatants) to more than $10\,g\,l^{-1}$. As a general guideline, when the concentration is greater than $1\,g\,l^{-1}$, the first step can be used to achieve a higher degree of purity. For a dilute solution, concentration will be achieved by using chromatographic media with capacity for high flow rates. The column will be short and wide; it will be operated at a pH and flow rate at which the product will bind, and the majority of the contaminants pass through. Alternatively, ultrafiltration may be used for concentration.

TOTAL PROTEIN AND NUCLEIC ACID CONCENTRATION

If the protein and/or nucleic acid content of the starting material is very high, the solution may be quite viscous. If the viscosity is greater than approximately 4 cP, it may be necessary to reduce the viscosity of the feed to prevent an increase in the back pressure of the column, and hence a decrease in flow rate with concomitant lowered throughput at a given pressure. Chromatographic bed compression may occur with insufficiently rigid media. Alternatively, larger particle size separation media may be used.

The ratio of total protein to product can vary between 1 and 10,[4] depending upon the host. If the ratio is very high, it is advisable to adsorb the product, rather than the contaminants, during the first step. When the ratio is close to one, well-defined contaminants can be adsorbed instead of the product. It will be necessary to test the capacity and specificity of the separation media to see which approach gives the best result.

CONDUCTIVITY

Generally, a conductivity of greater than $5\,\mathrm{mS\,cm^{-1}}$ at neutral pH will cause capacity problems during ion exchange because the small charged molecules in the feed will bind to the exchanger and limit the amount of binding sites available for protein binding. If the protein can retain its biological activity at a pH far from neutrality and its isoelectric point, it is sometimes possible to keep the protein highly charged so that it can compete with small ionic molecules and bind effectively to the exchangers.

Tissue preparations, plasma, normal body fluids and complex animal cell culture media give conductivities of approximately $15\,\mathrm{mS\,cm^{-1}}$; bacterial cultures can have a conductivity up to $30\,\mathrm{mS\,cm^{-1}}$; yeast cultures usually have a conductivity about $20\,\mathrm{mS\,cm^{-1}}$ (*Table 5*). It is, therefore, probable that if ion exchange chromatography is used as a first step, the sample will have to be conditioned. This can be done by diluting the sample or by 'desalting' by gel filtration or diafiltration. If the product is present in a low concentration (e.g. about $10\,\mathrm{mg\,l^{-1}}$), it may be useful to further dilute the sample to achieve a conductivity of less than $5\,\mathrm{mS\,cm^{-1}}$ and then pass it through an ion exchanger at a very high linear velocity (between 200 and $400\,\mathrm{cm\,h^{-1}}$) as long as the sample binds efficiently. (Since the kinetics of binding of large molecules vary considerably—depending on net charge and charge distribution—the optimal flow rate for each protein has to be determined empirically.) This approach was taken in the development of a process for a large scale purification of *Staphylococcus* enterotoxin B.[1] The

Table 5. Approximate osmolality values for different culture conditions.

Source	Osmolality (mOsm)
Mammalian cells	300
Bacteria:	
internal	220
external	500
Yeast:	
internal	220
external	500

alternative approach, desalting, is most useful if the product is present at high concentrations ($>1\ g\ l^{-1}$). In this case, gel filtration media with exclusion limits of approximately 5000 daltons can be used at bed heights up to 60 cm. Linear velocities up to $250\ cm\ h^{-1}$ and sample loadings up to 30% of the total bed volume can be used. Conditioning by gel filtration has the added advantage of removing small charged contaminants that can decrease the binding capacity of an ion exchanger for the product. In contrast to other modes of gel filtration, the sample dilution in desalting is only on the order of 1.1-1.5. This approach was taken in the purification of human superoxide dismutase produced in yeast.[2]

pH

Starting material from tissue and other natural sources will have differing pH values, depending on the source. For example, in the purification of human superoxide dismutase, the pH of the clarified yeast lysate was 5.6. *Staphylococcus aureus* cell culture supernatant from which enterotoxin was purified had a pH of 7.3.

pH adjustments will need to be made to preserve biological activity and to prepare the sample for chromatography. pH adjustments can be made that will allow the binding of the product, not the contaminants (or vice versa), in ion exchange or hydrophobic chromatography.

CONTAMINANTS

Lipids, polysaccharides, nucleic acids, cell fragments and debris, and small molecules (e.g. amino acids, pigments, additives from cell culture such as the indicator phenol red) can interfere with chromatographic steps. Non-specific adsorption can occur when contaminants accumulate on the chromatographic media. Chromatographic beds may become clogged by lipids and cell debris. In the case of proteins produced in the periplasmic space, lipids released from the cell membrane during cell disruption can interfere with subsequent chromatographic steps. In ascites fluid the lipid content is often quite high, and delipidation is recommended prior to chromatography to avoid clogging. Contaminants can also compete with the product for column capacity. In some cases, a desalting step carried out at high flow rates may serve to remove lipids and small molecules such as pigments. Alternatively, the first chromatographic step might be one that allows the product to bind while the contaminants pass through, e.g. affinity or ion exchange. For example, if phenol red is present, it will bind to an anion exchanger and take up valuable capacity. Using a cation exchanger as a first step to bind the product prevents this loss of column capacity.

In summary, whether the product is naturally occurring or obtained by recombinant DNA or hybridoma technology, it is necessary to consider the characteristics of the starting material to determine the optimal isolation step. Following isolation of the product, analysis of the feed will lead to selection of the most appropriate approach for sample preparation for the initial purification step. An efficient beginning is part of a cost-effective purification process.

REFERENCES

1. Johansson, H. O. J., Pettersson, N. T. and Bergloff, J. H. Development of a process for large scale purification of *S. aureus* cell culture supernatants. Third Preparative-scale Liquid Chromatography Symposium, 4-5 May 1987.
2. Daniels, A. I., Pettersson, N. T., Berglof, J. H. and Scandella, C. Purification of recombinant human superoxide dismutase produced in yeast: a process development study. Third Preparative-scale Liquid Chromatography Symposium, 4-5 May 1987.

6 Optimization

As discussed in the previous chapter, the chromatographic techniques selected will be dependent on the feed characteristics. Method scouting, i.e. the screening of a number of separation media and buffers, will lead to the selection of the type of media, ligand and degree of substitution—in other words, the selection of the most appropriate chemistries (see *Table 6*). Input data on the scale of operation can then be used to determine the physics, i.e. particle size and rigidity. This two-step process will enable the best process media to be identified.

Following method scouting, each step should be optimized. In contrast to analytical chromatography, in which the accent is on resolution and analysis time, the amount of product produced per unit time and the degree of purity are of the utmost importance in preparative chromatography.[1] In large scale chromatography, the critical parameters that must be optimized for each step are volume load, mass load, flow rate, resolution (determined by selectivity and efficiency), recovery, purity, bed configuration and cost (see *Table 7*).

Table 6. Optimization: selection of process media.

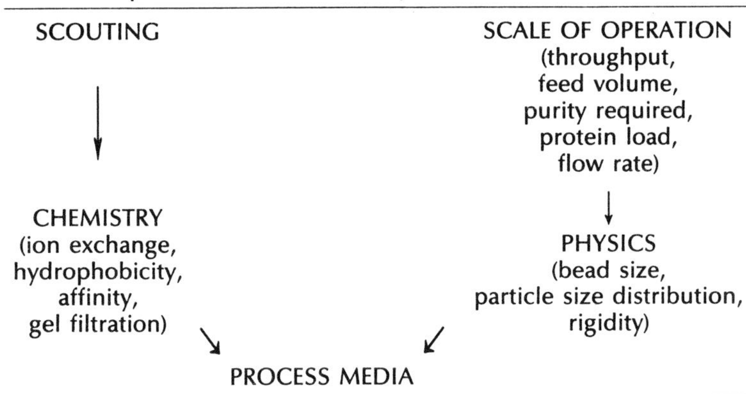

Table 7. Optimization: characterization of critical parameters.

Volume load
Mass load
Flow rate
Resolution
Recovery
Bed configuration
Purity
Cost

Volume load, mass load and flow rate will determine throughput. The chemistries and physics which control selectivity and efficiency, and thus resolution, are optimized to achieve maximum throughput and recovery (calculated as kilograms of adequately purified and biologically active product per hour).[2] The factors crucial to achieving optimal throughput are flow resistance and the dynamic capacity of the separation media. Dynamic capacity, i.e. the binding capacity of chromatographic adsorbents under actual working conditions, for a particular solute is dependent on several factors: matrix composition and pore structure; particle diameter and particle size distribution profile; product molecular weight and solubility; kinetics of the adsorption and desorption processes; bulk-, film-, and gel-diffusion constants of the product; the event of possible competitive binding and displacement effects of other proteins present in the sample feed solution.[2]

The dynamic capacity can be checked initially in a small batch procedure by measuring the disappearance of the sample from a stirred mixture of sample and adsorbent. A small column packed with the adsorbent can then be used to establish the process parameters. Concentration of product in the effluent can be plotted versus volume or time at different linear flow rates.[3] Two recent studies,[4,5] describe the use of frontal analysis for determining dynamic capacities at different flow rates. An abrupt accumulation of sample in the eluate indicates that the column is being used to its full capacity. The preferred linear flow rate will be determined not only by the capacity of the column at a given flow rate but also by yield and process requirements.

Process chromatography media should have the porosity and degree of substitution necessary to give maximum dynamic capacity. Affinity media should be easily derivatized. Since most biological molecules prefer an aqueous environment, the gels should be hydrophilic. The media should not bind molecules nonspecifically, so that good recoveries are obtained, and the media should give the necessary degree of resolution. Matrix requirements are summarized in *Table 8.*

To achieve maximum resolution on a laboratory scale, small particle size media are used. Preferred analytical HPLC media particle

Table 8. Matrix properties of media for large scale chromatrography.

Inert
Hydrophilic
Non-biodegradable
Chemically and physically resistant
Insoluble
Easily derivatized
Easily packed in small and large columns
Macroporous

sizes have been getting smaller, from 10 to 40 μ packings used most frequently in the 1960s, to the 5-10 μ packings of today. Smaller particles make columns more efficient, but in the purification of gram (or larger) quantities, the price may be prohibitive. Not only are smaller packings usually more expensive but, in most cases, the pressure drop is high, the equipment required to run the column is more costly and the sample load capacity may be too low to achieve sufficient throughput. The question of what efficiency is required for preparative HPLC has been addressed by Guiochon and Colin.[1] In some cases, relatively large particle sizes can give sufficient resolution. For example, porous silica ion exchangers with 30-50 μ diameter particles have been used to purify ovalbumin with the same resolution as that found on a 6 μ packing.[6] Sitrin *et al.*[7] have found that fewer than 300 theoretical plates are required for the purification of polar glycopeptide antibiotics by RPC. This permitted the use of less costly 40-60 μ particles for the purification of multi-gram quantities. On the other hand, for more difficult separations, 2000 plates were required, and 5-10 μ particles used. In a discussion of high performance gel filtration, ion exchange, hydrophobic interaction and reversed-phase by Johnson,[8] it is recommended that "the largest particle size which will provide the required separation should be used preparatively to minimize back pressure and capital expense." Janson and Hedman have concluded: "It is clear that above a certain level very little is gained in production rate by increasing the column efficiency, i.e. the plate number, *N*, by reducing the particle size. On the other hand, a dramatic increase in throughput can be achieved by increasing the selectivity of the chromatographic process. Increased selectivity can be obtained by further optimizing the conditions in ion exchange chromatography by changing the ligand in affinity chromatography, or by chosing another gel pore distribution in gel filtration."[2]

ION EXCHANGE

Important factors to consider when optimizing ion exchange chromatography are: sample load, flow rate, and gradient volume.

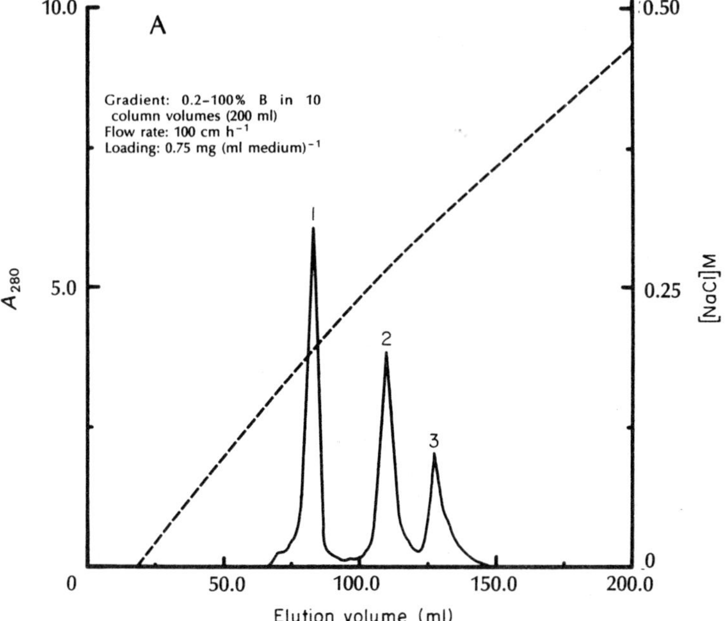

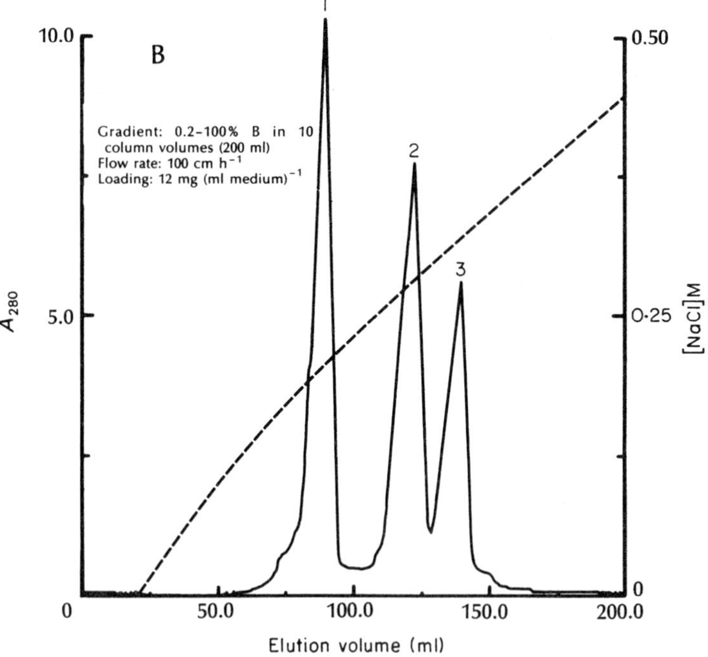

Figure 4 (See page 29 for caption)

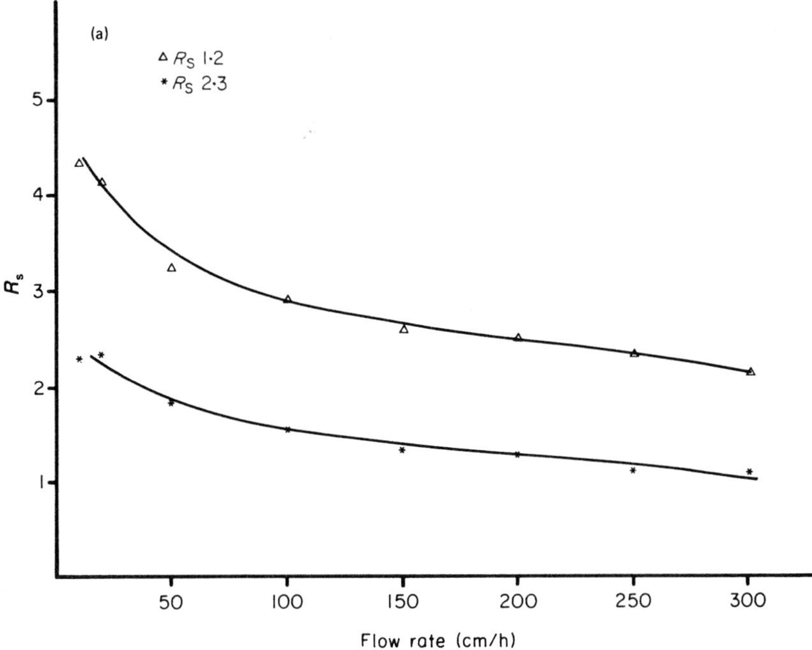

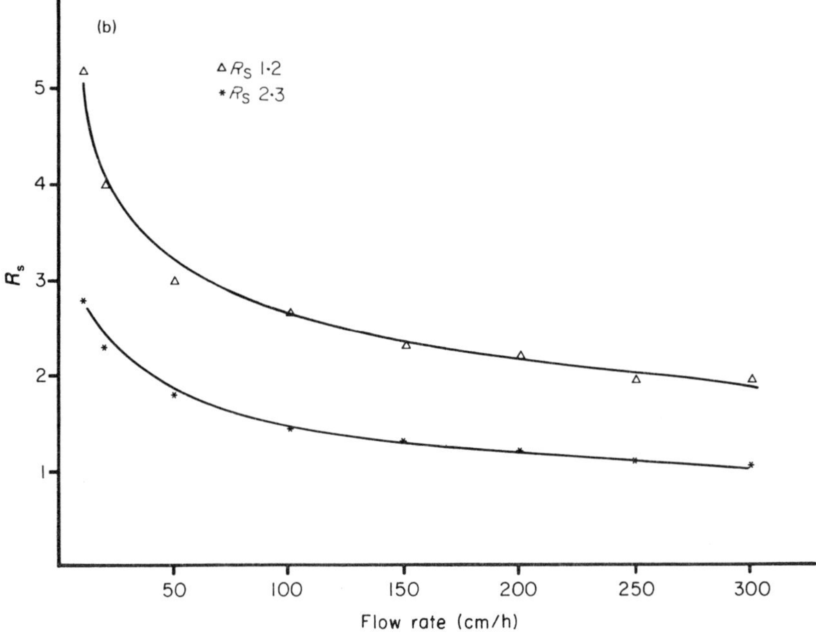

Figure 4 (See page 29 for caption)

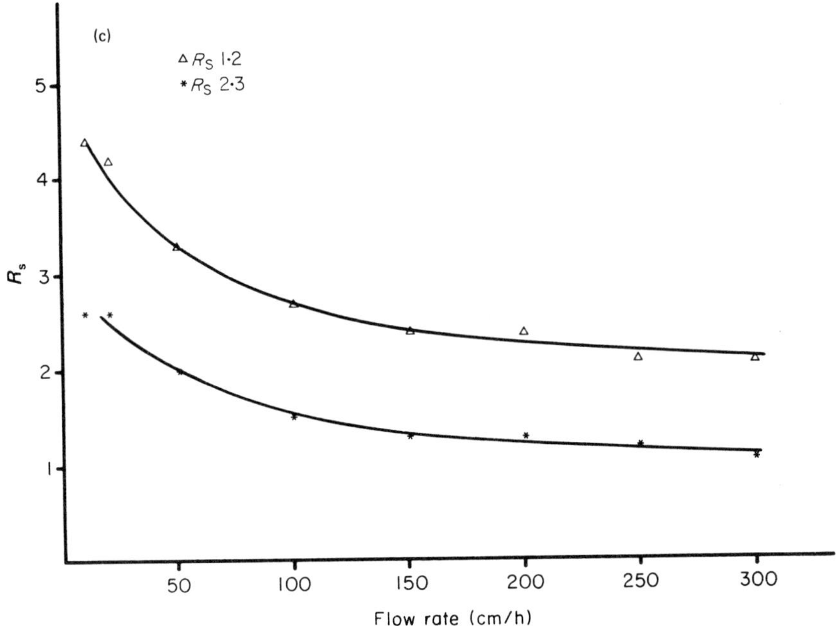

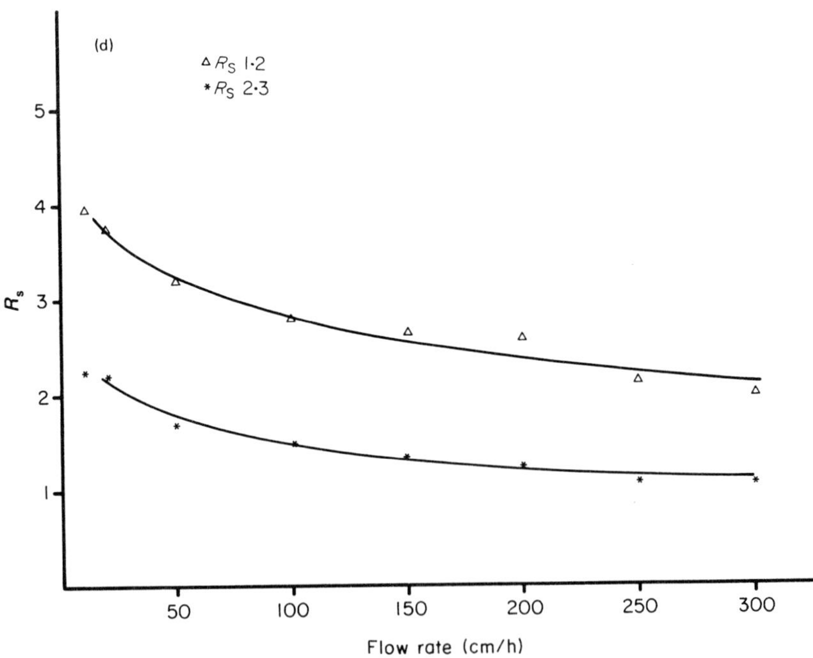

Figure 4 (See page 29 for caption)

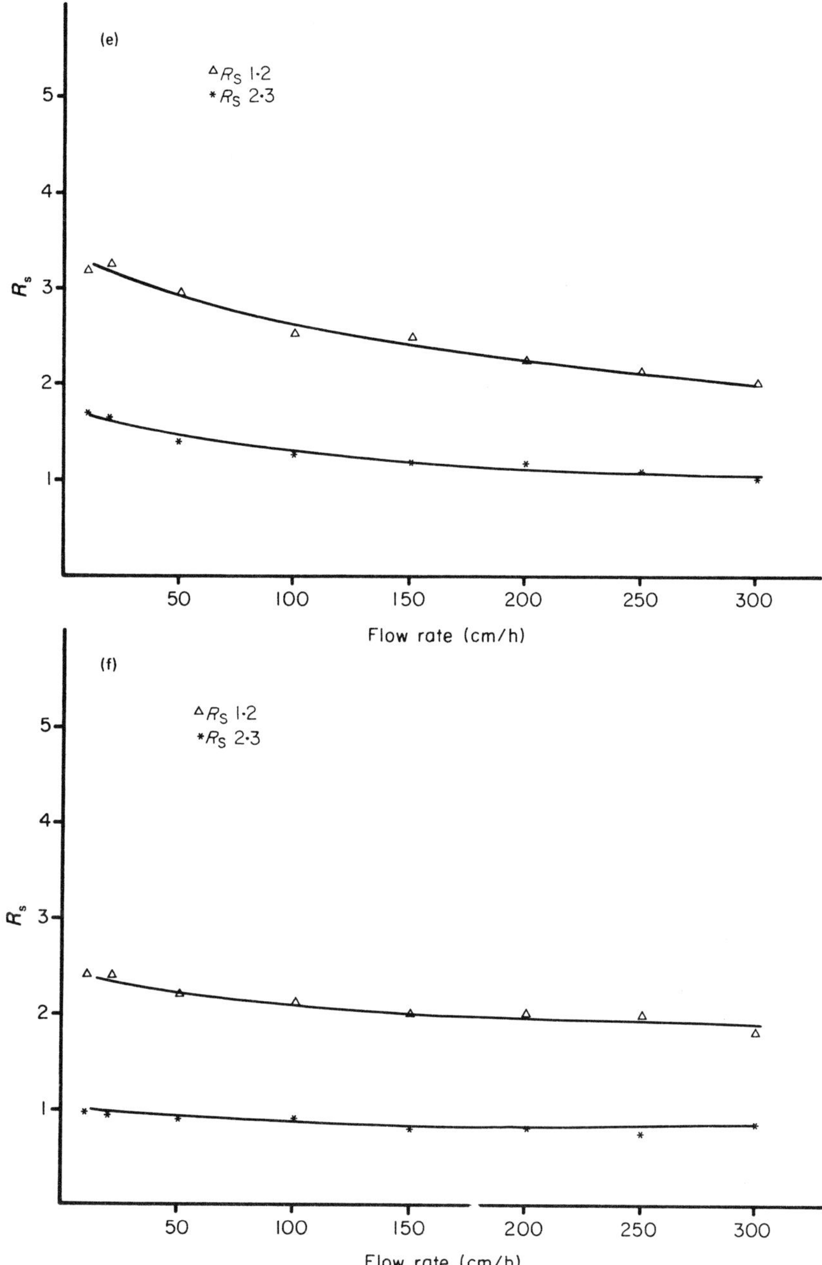

Figure 4. The effects of sample load and flow rate on resolution in ion exchange chromatography. △ R_s 1·2 indicates the R_s values for peaks 1 and 2; *R_s 2·3 indicates R_s values for peaks 2 and 3. (An R_s value of 1.2 indicates baseline resolution.) Medium, S Sepharose High Performance; column, HR 16/10; buffer A, 20 mM HCOOH, pH 4.0; buffer B, 20 mM HCOOH, 0.5 M NaCl; gradient, 0.2–100% B; sample, wheat germ isolectin 15 mg ml^{-1}. Baseline loads: (a) 0.38 mg (ml medium)$^{-1}$; (b) 0.75 mg (ml medium)$^{-1}$; (c) 1.5 mg (ml medium)$^{-1}$; (d) 3.0 mg (ml medium); (e) 6.0 mg (ml medium)$^{-1}$; (f) 12 mg (ml medium)$^{-1}$. Parts A and B show examples of actual chromatograms used for evaluation.

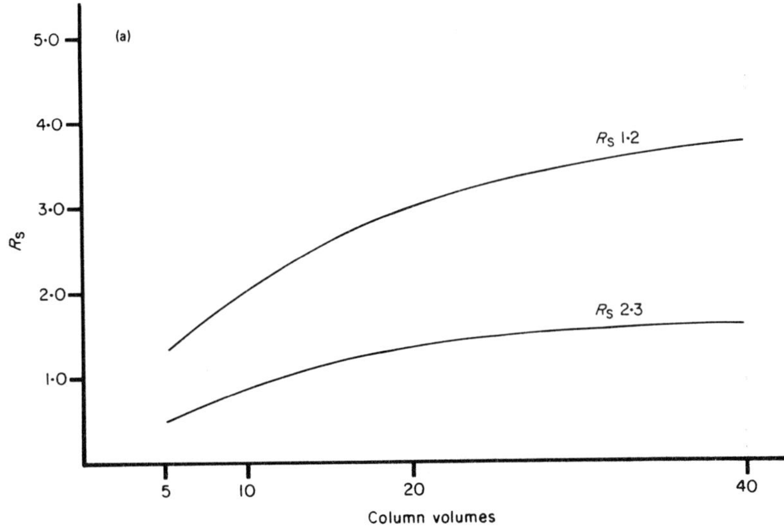

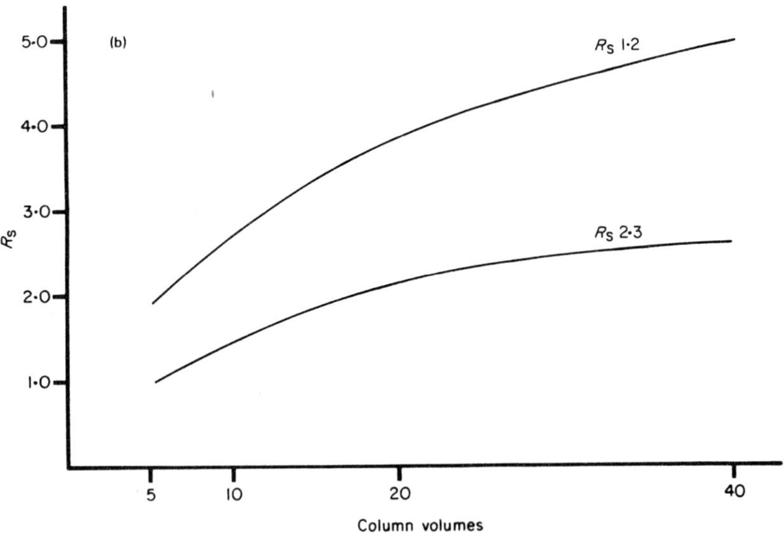

Figure 5 (See page 31 for caption)

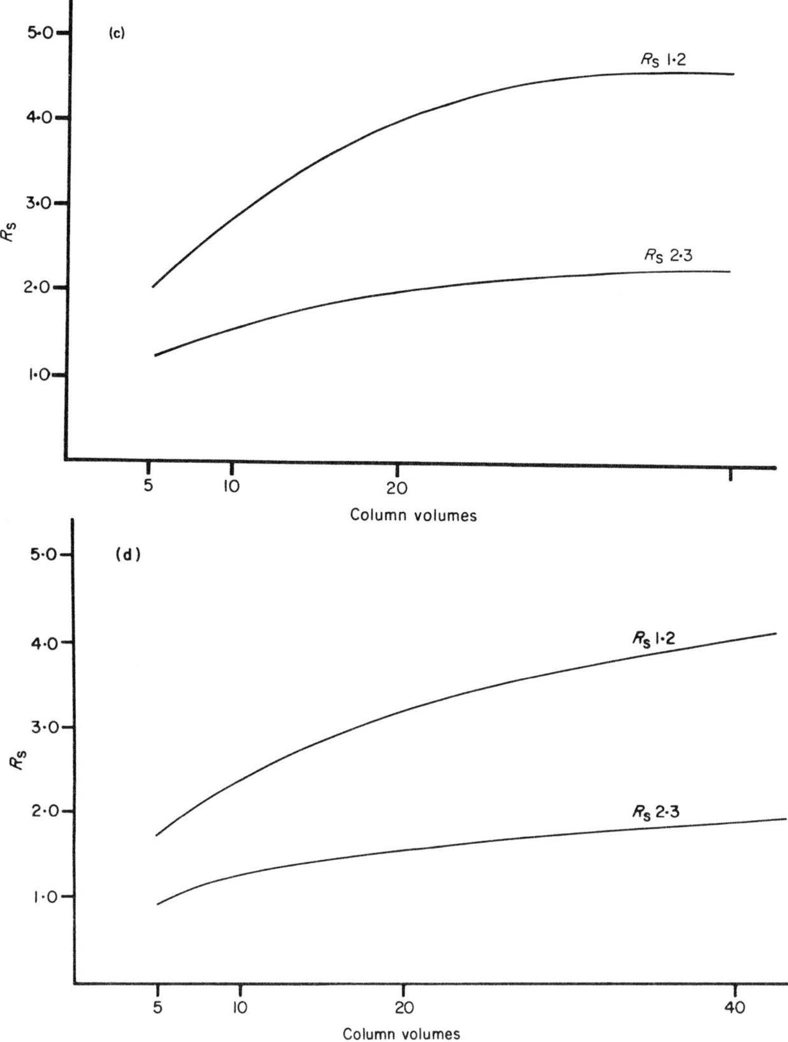

Figure 5. The effect of gradient volume and flow rate on resolution of wheat germ lectin by ion exchange chromatography. Medium, S Sepharose High Performance; column, HR 16/10; buffer A, 20 mM HCOOH, pH 4.0; buffer B, 20 mM HCOOH, 0.5 M NaCl; gradient, 0.2–100% B; sample, wheat germ isolectin 15 mg ml^{-1}. (a) Sample load, 12 mg (ml medium)$^{-1}$; flow rate, 100 cm $^{-1}$. (b) Sample load, 3 mg (ml medium)$^{-1}$; flow rate, 100 cm $^{-1}$. (c) Sample load, 0.75 mg (ml medium)$^{-1}$; flow rate, 100 cm h^{-1}. (d) Sample load, 0.75 mg (ml medium)$^{-1}$; flow rate, 200 cm h^{-1}.

All of these factors are somewhat interrelated. *Figure 4* shows that at a low sample loading, resolution is improved significantly by decreasing the flow rate, while at a high sample loading, resolution is improved only slightly by decreasing the flow rate. This means that under production conditions, where maximum sample load is applied to achieve maximum throughput, the flow rate is limited primarily by the rigidity of chromatography media and by system constraints. In addition, the selectivity afforded by the ion exchanger for the product should be considered. For example, it can be seen in *Figure 4* that in the purification of wheat germ isolectin, peaks 1 and 2 are resolved even at the highest sample load and at flow rates up to 300 cm h^{-1},[9] whereas peaks 2 and 3 are resolved only at sample loads up to 6.0 mg (ml gel)$^{-1}$ at a flow rate of 160 cm h^{-1}.

In process chromatography, the best results will be obtained by using the maximum flow with the gradient volume that provides the best resolution. This is shown in the following example.

Figure 5 shows that resolution increases with increasing gradient volume. The pattern is similar for the resolution between peaks 1 and 2 and between peaks 2 and 3.

As the sample load is increased, increased gradient volumes are required to give the same resolution. To resolve peaks 2 and 3 with an R_s value of 1.2 at a flow rate of 100 cm h^{-1}, gradient volumes of 5 column volumes were required for the sample load of 0.75 mg, 9.5 column volumes for the load of 3 mg, and 16.5 column volumes for the load of 12 mg.

As the flow rate is increased, an increased gradient volume will be required to achieve the same resolution. For example, when the flow rate was increased from 100 to 200 cm h^{-1}, a gradient volume of 8.5 column volumes (instead of 5) was required to resolve peaks 2 and 3 at a 0.75 mg sample load (*Figure 5c,d*). As the resolution increases with increased gradient volume, however, a dilution of each

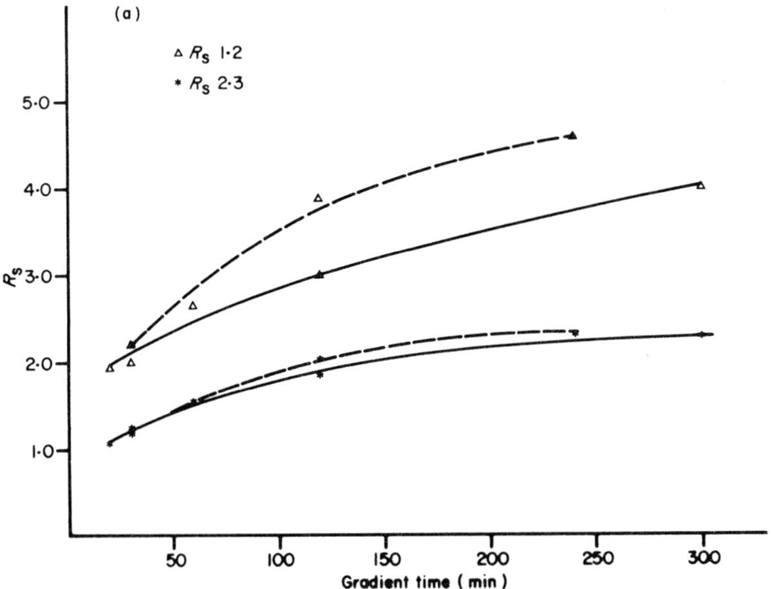

Figure 6 (See page 33 for caption)

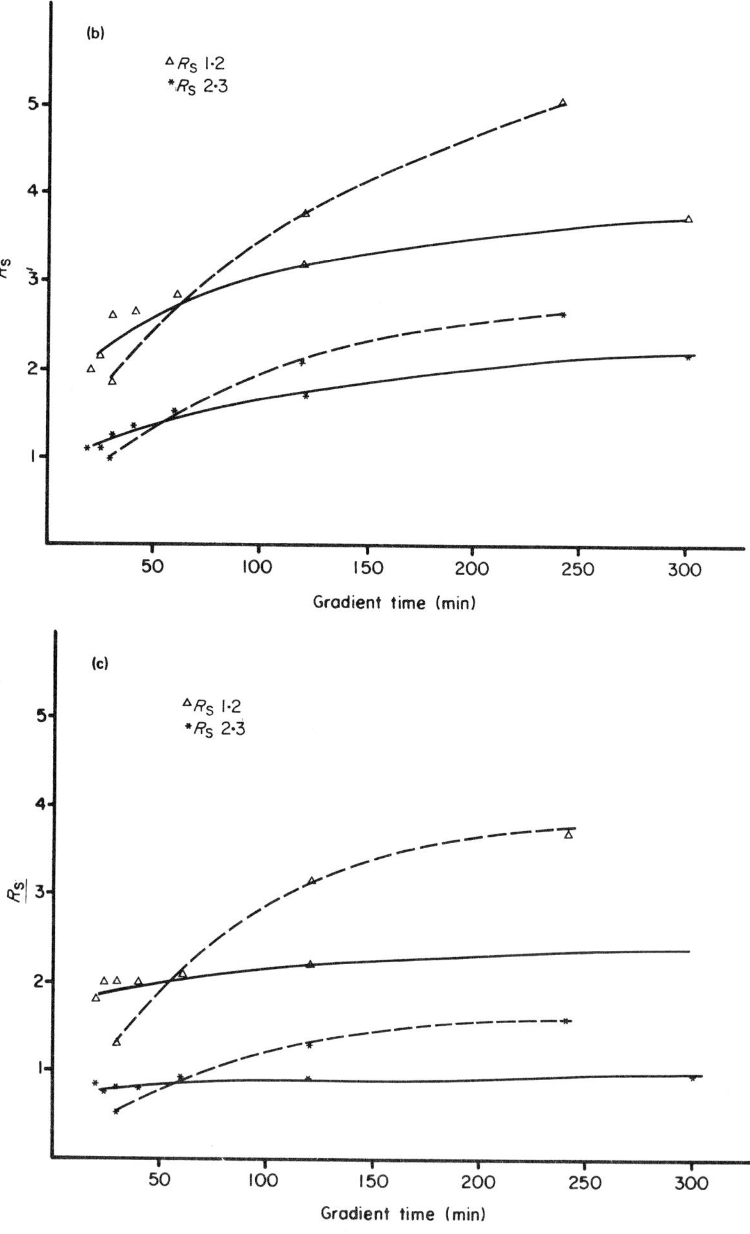

Figure 6. The effect of sample load and gradient time on the resolution of wheat germ isolectin by ion exchange chromatography. Key: ------, constant flow rate and variable gradient volume; ——, constant gradient volume and variable flow rate. Medium, S Sepharose High Performance; column, HR 16/10; gradient, 0.2–100% B; sample, wheat germ isolectin 15 mg ml^{-1}. (a) Loading, 0.75 mg (ml medium)$^{-1}$; flow rate, 100 cm h^{-1}, gradient 5–40 column volumes; gradient, 10 column volumes, flow rate 20–300 cm h^{-1}. (b) Loading, 3 mg (ml medium)$^{-1}$; flow rate, 100 cm h^{-1}, gradient 5–40 column volumes; gradient, 10 column volumes, flow rate 20–300 cm h^{-1}. (c) Loading 12 mg (ml medium)$^{-1}$; flow rate, 100 cm h^{-1}, gradient 5–40 column volumes; gradient, 10 column volumes, flow rate 20–300 cm h^{-1}.

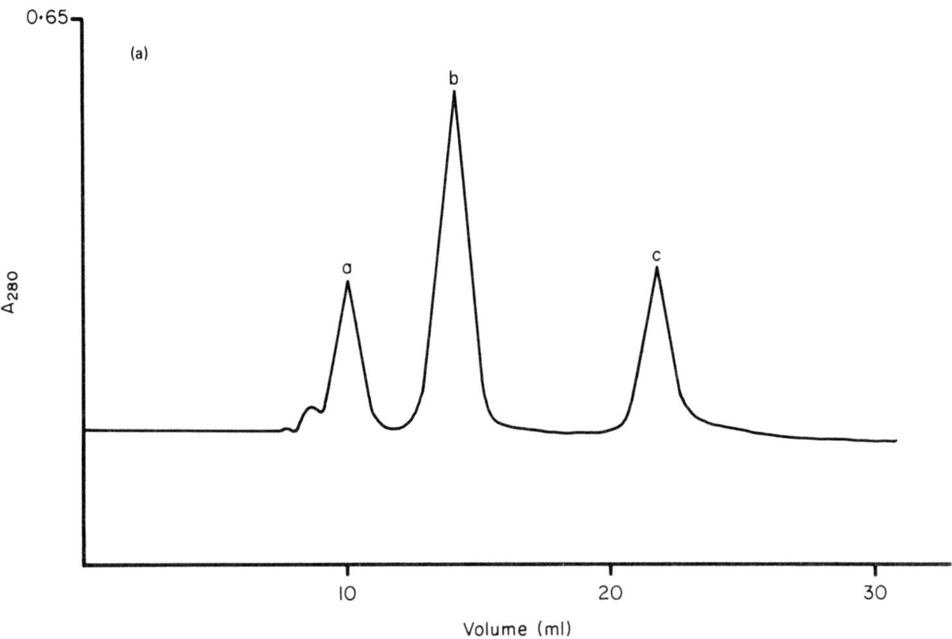

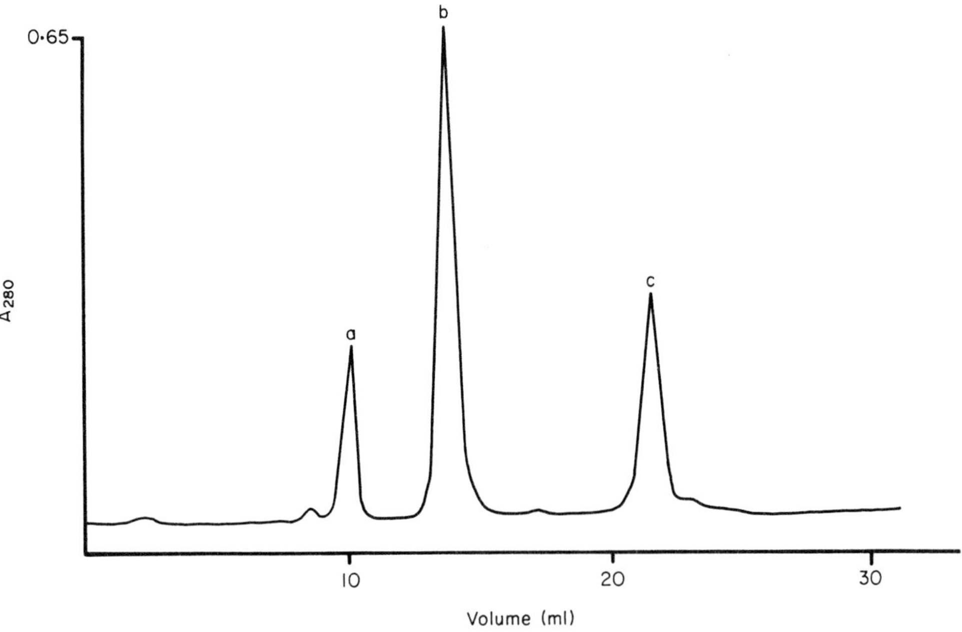

Figure 7 (See page 35 for caption)

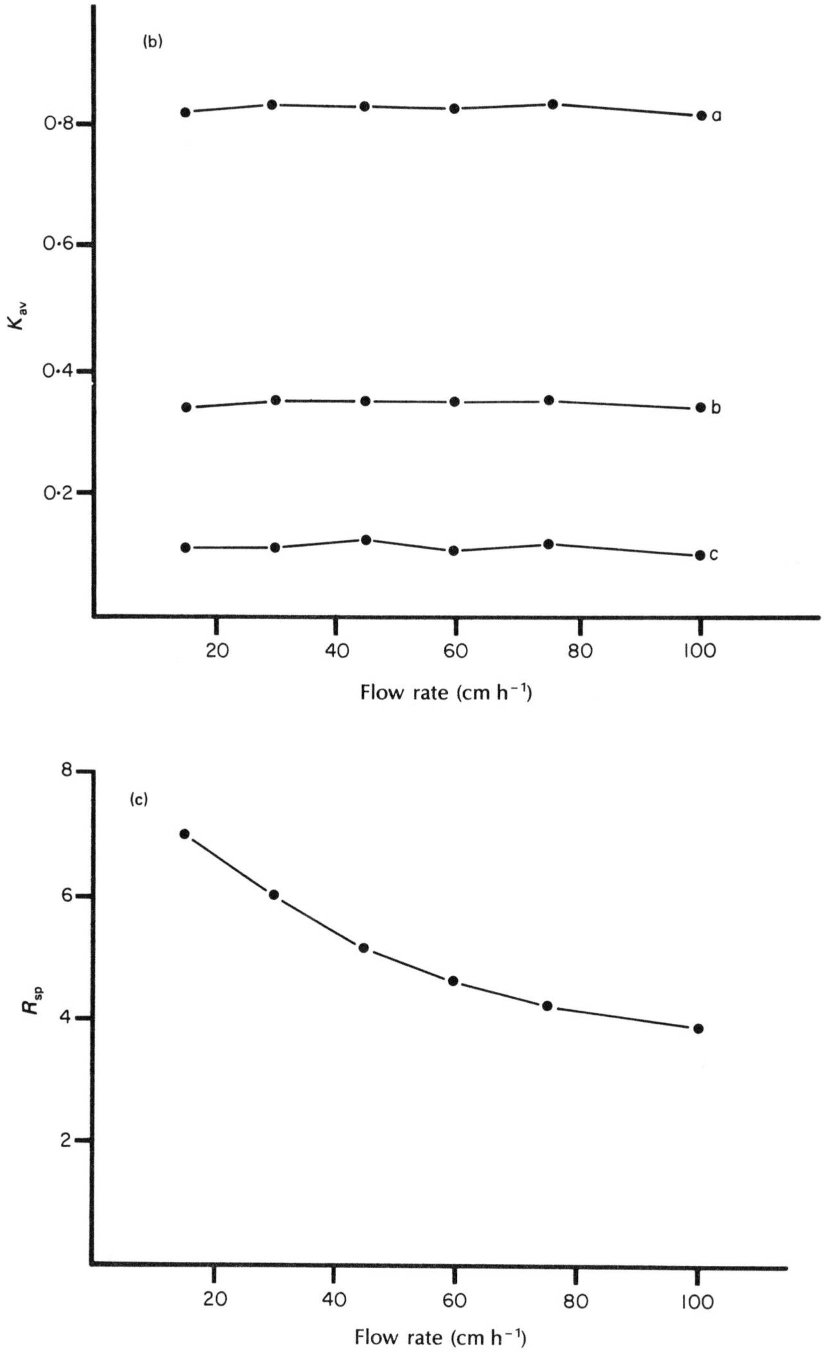

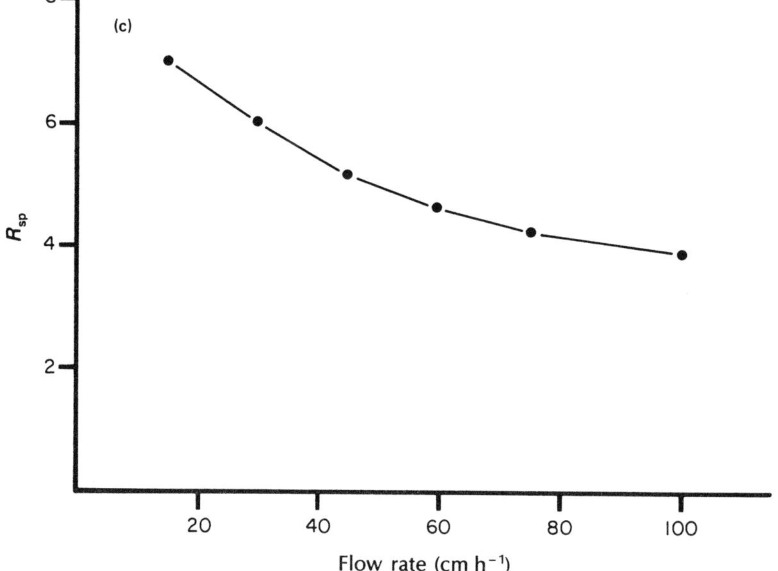

Figure 7. The effect of flow rate on gel filtration as shown in representative chromatograms: (a) BSA, (b) α-chymotrypsin, and (c) vitamin B_{12}. Medium, Superdex® 75; column, HR 10/30; eluent, 0.05 M Na_3PO_4, pH 7.2 + 0.1 M NaCl; V_t sample volume: 0.5%. (Work from Pharmacia LKB Biotechnology AB, Jonson, A.-K.)

peak occurs. Fortunately, the peak volume increases at a slower rate than the gradient volume.[10]

Figure 6 shows that with a sample load of 0.75 mg per ml S Sepharose High Performance, resolution increases with time in a uniform manner regardless of whether the gradient is increased or the flow rate decreased. But with higher sample loads, from 3 to 12 mg per ml S Sepharose High Performance in this particular example, the resolution increases to a greater extent by increasing the gradient volume than by decreasing the flow rate. This effect is particularly pronounced in critical separation situations where $R_s = 1.2$ or less.

GEL FILTRATION

The important parameters to consider when optimizing gel filtration are: flow rate, efficiency and bed height. *Figure 7b* shows that flow rate has no effect on selectivity, measured by K_{av}. But resolution, measured by R_{sp} ($R_{sp} = R_s \times 1/\log(Mw_1/Mw_2)$), is dependent upon flow rate (see *Figure 7c*). The resolution of the standard proteins shown in *Figure 7* was more than satisfactory even at 100 cm h^{-1}, and it is conceivable that an even faster flow rate could have been used, provided that there were no system constraints.

Figures 8 and *9* show the effect of flow rate on efficiency, measured by plates (N) per meter in *Figure 8* and by reduced plate height in *Figure 9*. There is an optimal flow rate range for each chromatographic medium and for each sample. The flow rate at which optimal efficiency is obtained will vary depending on the molecular weight of the molecule of interest. Slow mass transfer of macromolecules leads to band

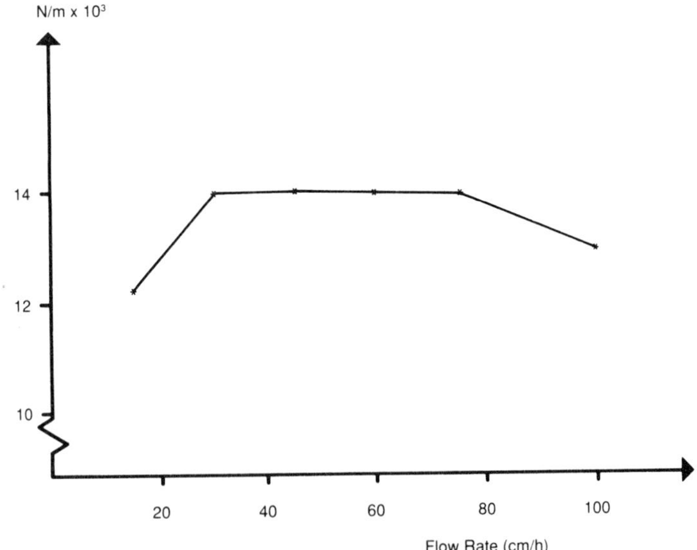

Figure 8. The effect of flow rate on efficiency measured by plates per meter (N/m). Medium, Superdex 75; sample, acetone; sample volume, 1% V_t.

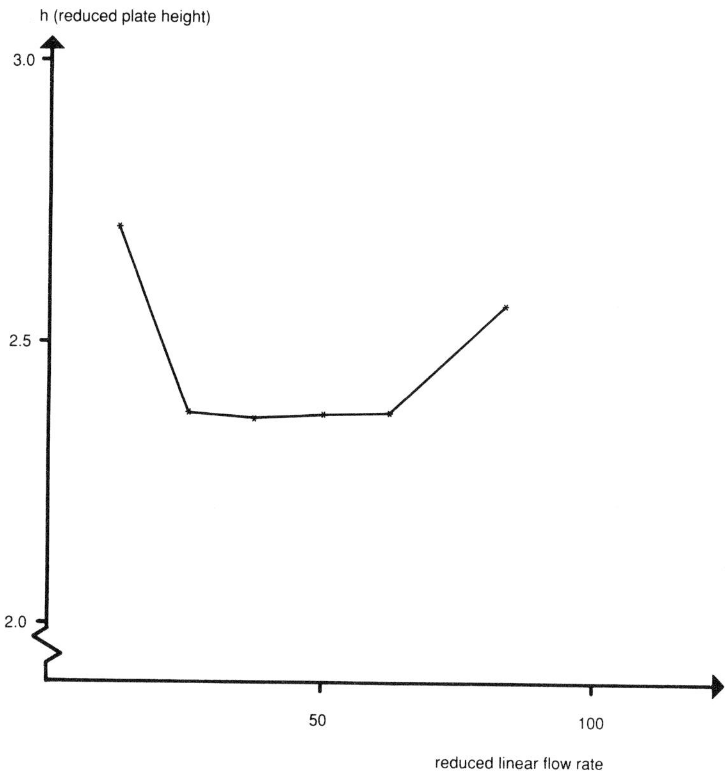

Figure 9. The effect of flow rate on efficiency measured by Knox plot.

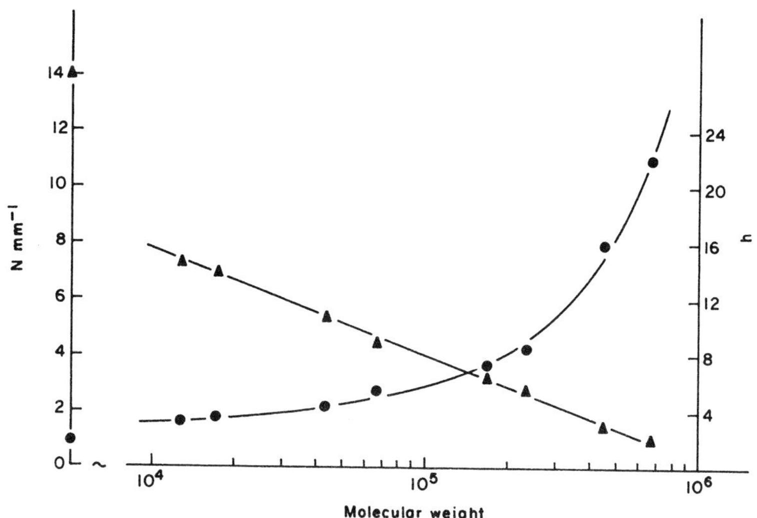

Figure 10. The number of plates (▲) and reduced plate height (●) as a function of sample molecular weight. Solutes used are AMP, cytochrome C, myoglobin, ovalbumin, BSA, γ-globulin, catalase, feritin and thyroglobulin. (From Hagel, L. and Andersson, T. *J. Chromatogr.* **285** (1984) 295–306.)

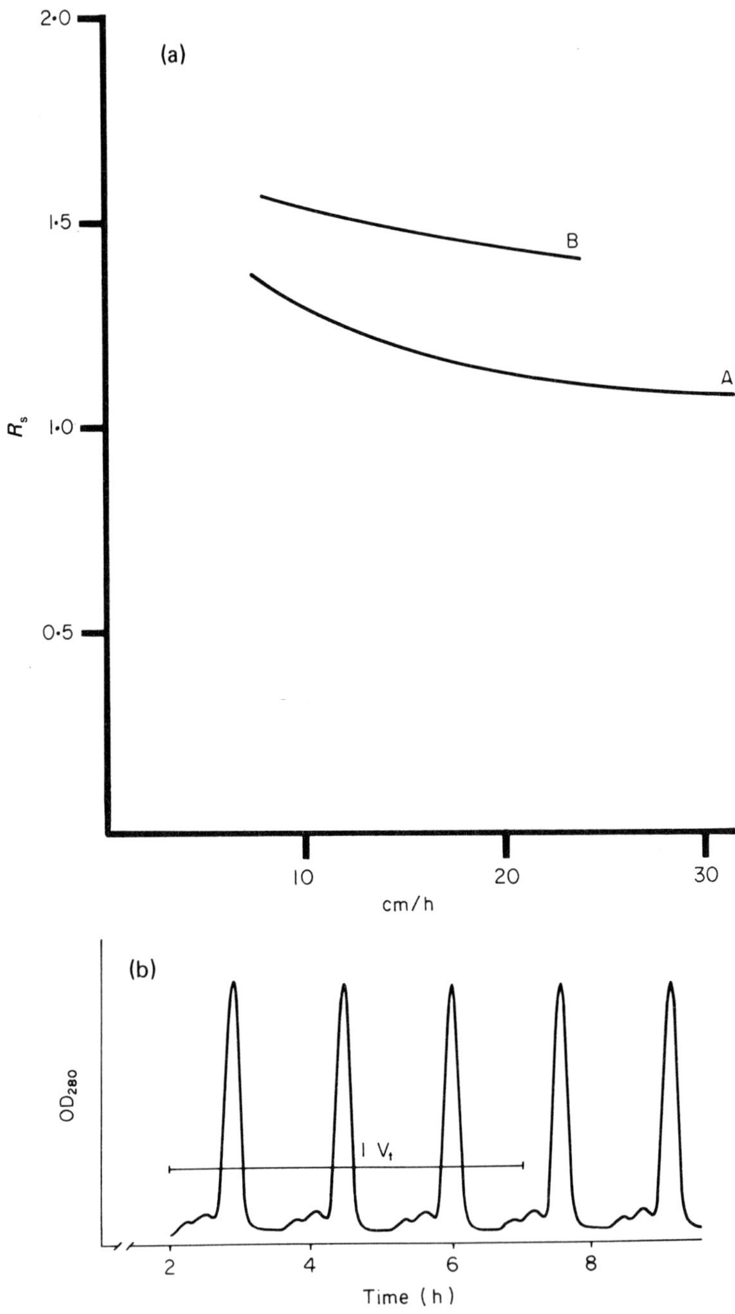

Figure 11. (a) Resolution (R_s) as a function of flow rate in columns of different bed heights: (A) 90 cm; (B) 180 cm. Medium, Sephacryl® S 200 High Resolution; sample, human albumin and IgG; sample volume, 4% V_t. (b) Gel filtration of albumin obtained from ion exchange chromatography in repeated cycles. Sample size, 4% V_t; concentration, 15 mg ml^{-1}; flow rate, 40 cm h^{-1}; bed height, 12–16 cm. The separations are spaced so three separations are achieved in an elution volume of one column volume (V_t), i.e. the removal of salt has not been considered. The distance corresponding to V_t is marked.

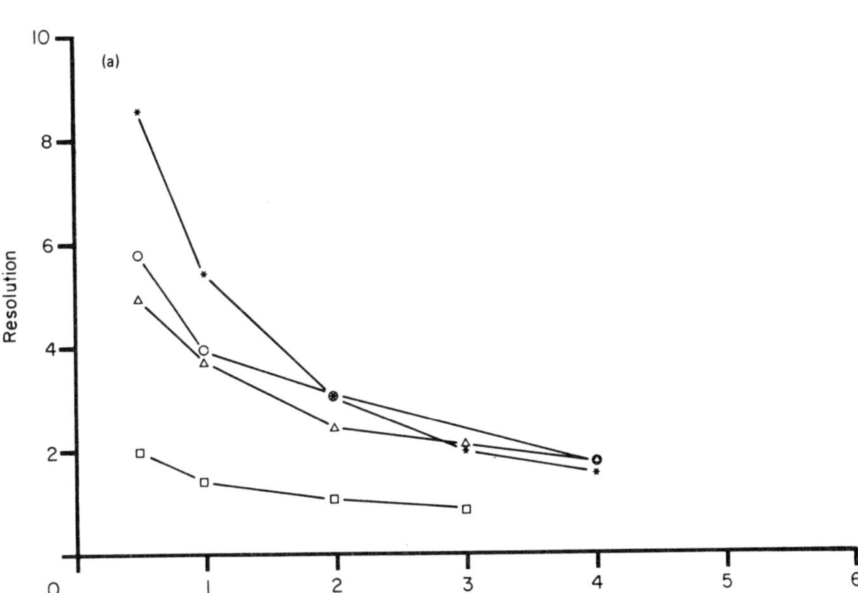

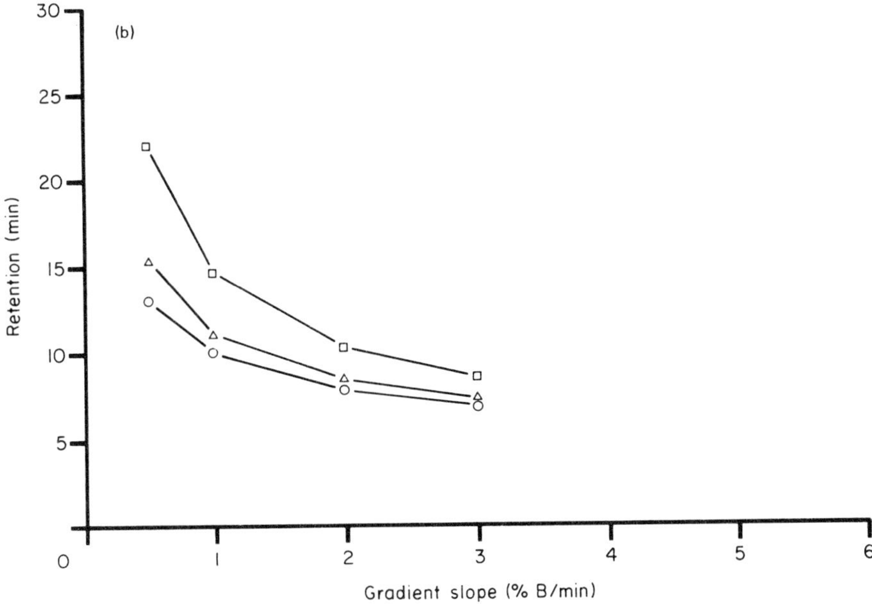

Figure 12. Effect of gradient slope on (a) resolution and (b) retention time. ∗, R_s between peptides (9) and (10); △, R_s between peptides (4) and (5); □, R_s between peptides (6) and (7); ○, R_s between peptides (1) and (2). Peptides: (1) tyr-calcitonin gene related peptide; (2) (gln^8)-LH-RH; (4) (des-arg^9)-bradykinin; (5) β-casomorphin; (6) tyr-bradykinin; (7) (asn^1, val^5, asn^9)-angiotensin I; (9) delicious peptide; (10) delta sleep inducing peptide. Column, Mono D® HR 5/20; buffer A, 50 mM H_3PO_4, pH 2 with NH_4OH; buffer B, 95% CH_3CN/H_2O; flow rate, 300 cm h^{-1}.

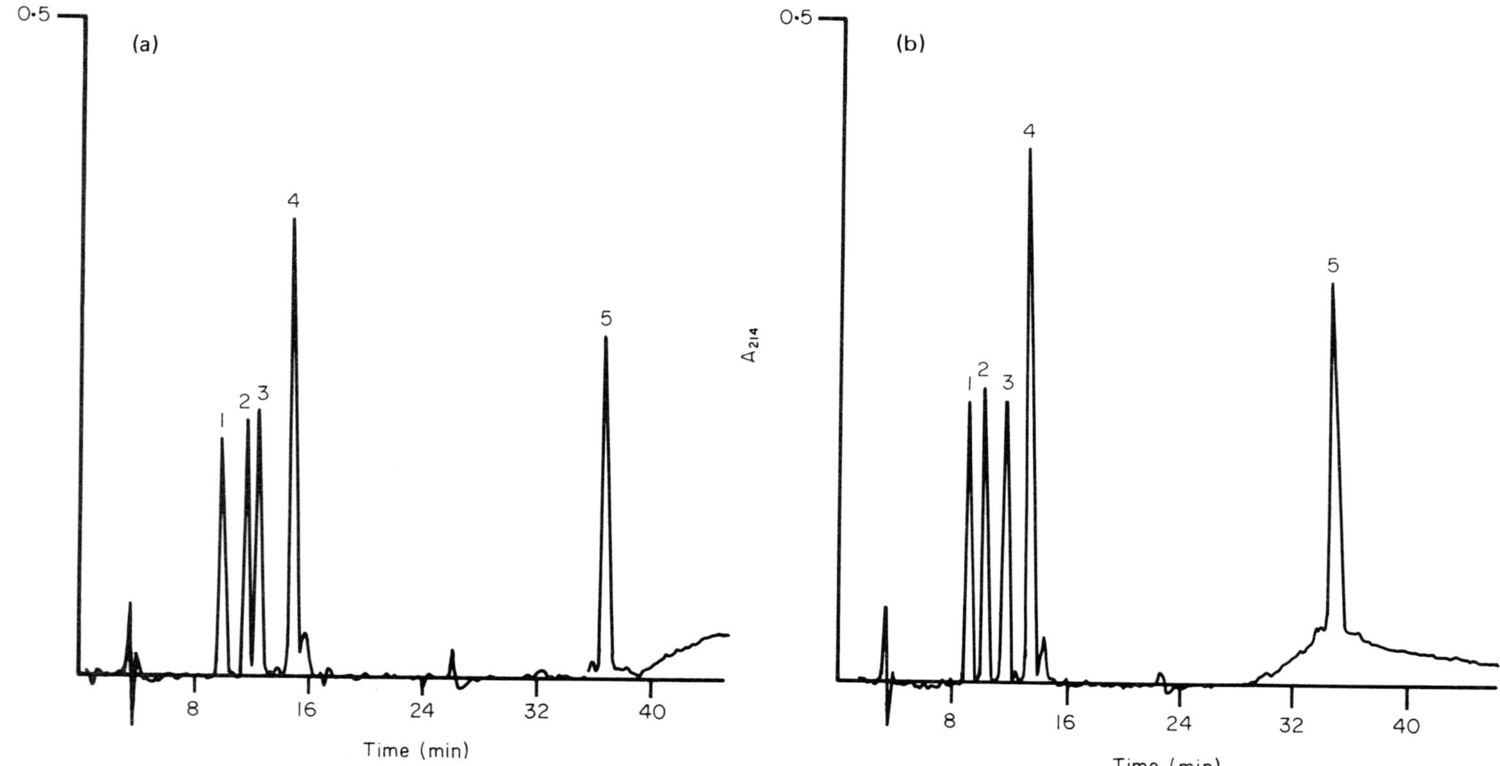

Figure 13. The effect of temperature on resolution in RPC: (a) 25°C and (b) 40°C. Column, Mono D® HR 5/20; buffer A, 50 mM H_3PO_4, pH 2 with NH_4OH; buffer B, 95% CH_3CN/H_2O; flow rate, 1.0 ml min^{-1}; chart speed, 0.25 cm h^{-1}. Samples: (1) oxytosin; (2) angiotensin II; (3) eledosin-related peptide; (4) angiotensin I; (5) glucagon.

broadening, and a lower flow rate may be required to compensate. The Van Deemter plot can be used for obtaining an approximate value of the column zone broadening.[11] The variation in column efficiency for proteins of different molecular weights is shown in *Figure 10.*

Further studies on the optimization of gel filtration have shown that optimal throughput can be achieved by using longer columns, provided the gel is rigid enough to withstand high flow rates. As seen in *Figure 11a*, longer columns give better resolution. It is possible to double the column length while doubling the sample volume and increasing the linear velocity to the point where sufficient resolution is achieved. Another way to increase throughput, especially in final purification steps, is to load more sample prior to elution of all components. This can be done since there will be very few components and their elution position known. An example of this is shown in *Figure 11b*.

Doubling the bed height will increase the resolution by about 40% ($R_s \propto \sqrt{L}$). Increasing the flow rate to the point where R_s is just achieved will increase productivity. However, the ability to pack a longer column with the same efficiency must be determined, and back pressure must be considered. Another alternative is to run two columns in series. In this case, the loss of resolution that occurs in the connections and between the two columns must be evaluated.

REVERSED PHASE CHROMATOGRAPHY

Important factors to consider when optimizing reversed phase chromatography (RPC) include: matrix stability, pH, gradient slope, temperature and flow rate. Dolan and Snyder have reviewed different ways to optimize RPC.[12] The stability of the support is important for process chromatography. Silica columns provide high resolution separations, but silica is chemically unstable and surface silanol groups become exposed when the matrix breaks down. The exposed silanol groups tend to cause peak trailing due to ion exchange effects. This trailing effect can be seen at a pH even as low as 3. The chemical instability of silica reduces the lifetime of the column. Polymeric RPC gels are stable to a wider pH range and may, therefore, provide increased column life. Even more important, the ability to vary pH permits greater selectivity to be obtained.

Since peptides contain mixtures of hydrophobic, hydrophilic, basic and acidic residues in any one sequence, and each residue changes in a different fashion with changes in pH, great selectivity can be achieved by altering pH.[13,14] During optimization, it may be useful to vary pH considerably to establish the best conditions. In this case, phosphoric acid is the best choice for the starting condition mobile phase due to its three pK_a values that cover the pH range 2-12.

Other starting condition mobile phases include hexafluorobutyric acid (HFBA), triethylammonium phosphate (TEAP)[15] and trifluoroacetic acid (TFA). Retention times for peptides are generally greatest for HFBA and least with phosphoric acid. TFA, the most commonly used starting condition mobile phase, also tends to ion pair to the peptide. Selectivity of RPC separations can thus be varied by the acidic component used and its concentration.

Isocratic elution is frequently used for elution of small molecules in RPC, but for peptides gradient elution provides the best resolution. *Figure 12* shows the effect of gradient slope on resolution and retention time. At a flow rate of 300 cm h^{-1} with a particle size of 10 μ, resolution increases as the gradient slope is reduced from 4 to 0.5% B min^{-1} (*Figure 12a*). Retention also increases as the slope is reduced (*Figure 12b*).

Temperature also has a profound effect on resolution. *Figure 13* shows the improvement in resolution that was achieved by raising the temperature from 25 to 40°C in the separation of a mixture of neutral peptides.

The effect of flow rate on retention of a mixture of basic peptides is shown in *Figure 14*. When the gradient slope was held constant at 1% B min^{-1}, resolution increased as the flow rate was increased from 0.25 to 1.50 ml min^{-1}.

The effects of alkyl chain length and pore diameter on protein retention, recovery and resolution in RPC have been reviewed by Steffensen and Anderson.[16]

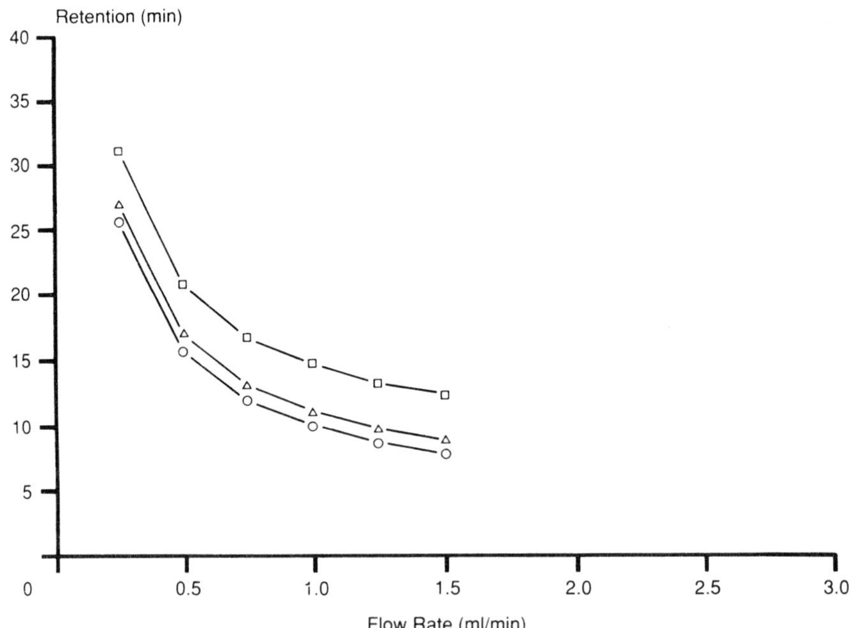

Figure 14. The effect of flow rate on retention in RPC. □, Substance P, free acid; △, (asn[1], val[5], asn[9])-angiotensin I; ○, try-bradykinen.

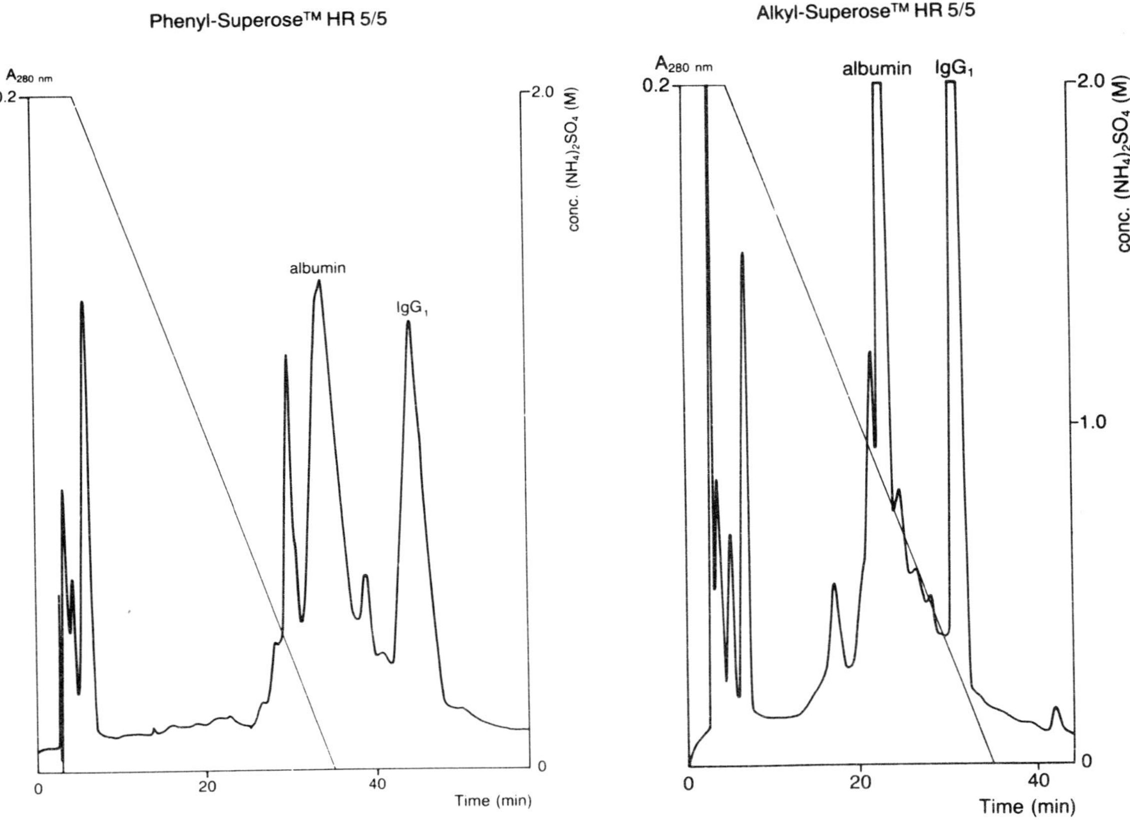

Figure 15. The effect of ligand on selectivity in HIC. Sample, 100 μl anti-CEA MAB (IgG_1) from mouse ascites fluid (diluted 1:2 with buffer A); flow rate, 0.5 ml min^{-1}; buffer A, 0.1 M phosphate, pH 7.0, 2.0 M ($(NH_4)_2SO_4$); buffer B, 0.1 M phosphate, pH 7.0, 0.02% NaN_3. (Work from Pharmacia LKB Biotechnology AB, Sparrman, M.)

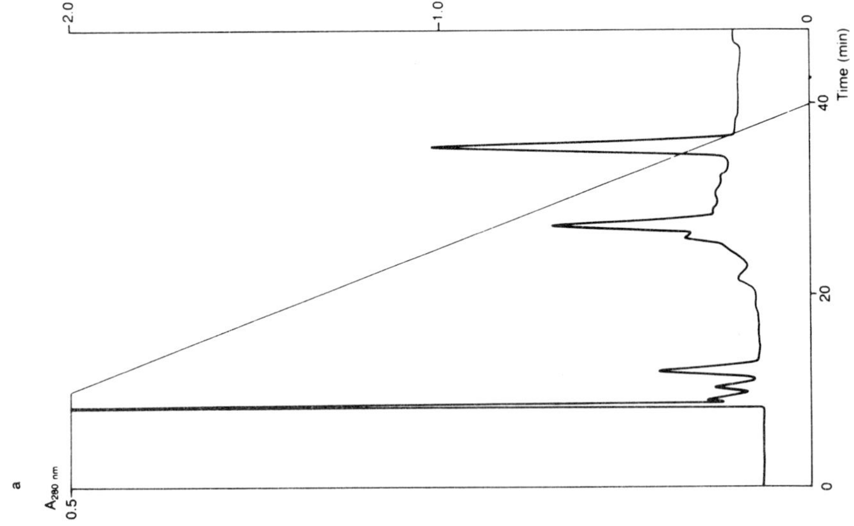
a
$A_{280\ nm}$
0.5
conc. $(NH_4)_2SO_4$ (M)
2.0
1.0
0
0
20
40
Time (min)

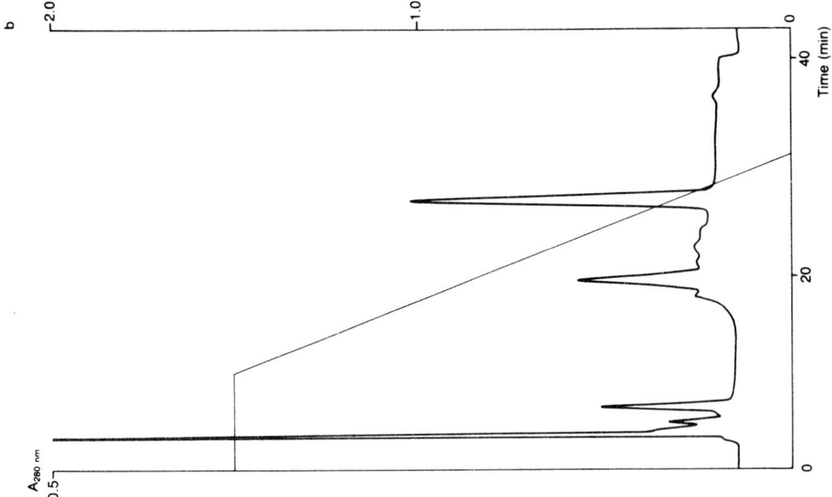
b
$A_{280\ nm}$
0.5
conc. $(NH_4)_2SO_4$ (M)
2.0
1.0
0
0
20
40
Time (min)

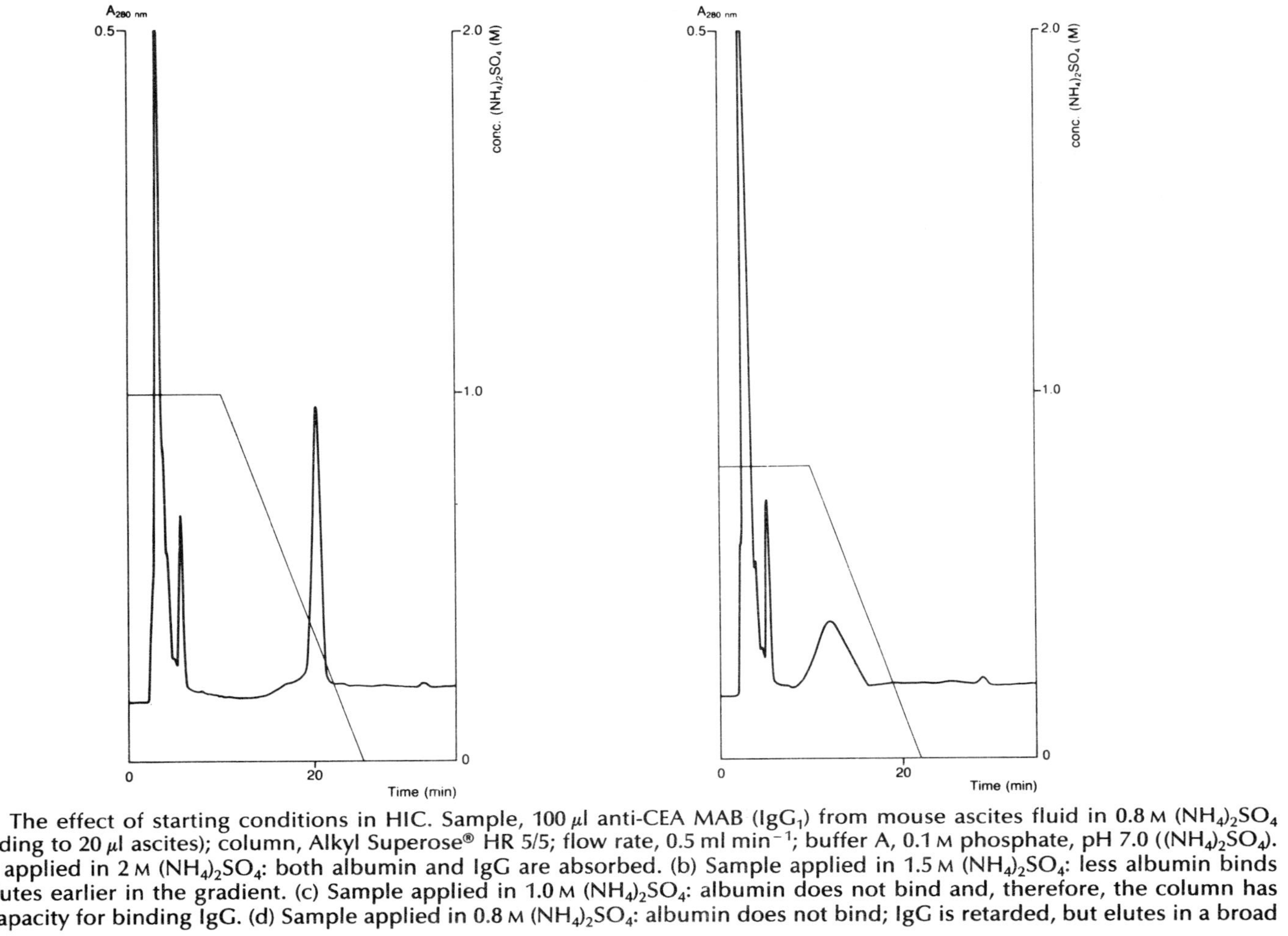

Figure 16. The effect of starting conditions in HIC. Sample, 100 μl anti-CEA MAB (IgG_1) from mouse ascites fluid in 0.8 M $(NH_4)_2SO_4$ (corresponding to 20 μl ascites); column, Alkyl Superose® HR 5/5; flow rate, 0.5 ml min^{-1}; buffer A, 0.1 M phosphate, pH 7.0 ($(NH_4)_2SO_4$). (a) Sample applied in 2 M $(NH_4)_2SO_4$: both albumin and IgG are absorbed. (b) Sample applied in 1.5 M $(NH_4)_2SO_4$: less albumin binds and IgG elutes earlier in the gradient. (c) Sample applied in 1.0 M $(NH_4)_2SO_4$: albumin does not bind and, therefore, the column has a greater capacity for binding IgG. (d) Sample applied in 0.8 M $(NH_4)_2SO_4$: albumin does not bind; IgG is retarded, but elutes in a broad peak.

HYDROPHOBIC INTERACTION CHROMATOGRAPHY

Important parameters to consider when optimizing hydrophobic interaction chromatography (HIC) include: hydrophobicity of the ligand, starting conditions, gradient elution, sample application and temperature. Hydrophobic interaction chromatography has been used to purify many proteins, including monoclonal antibodies. The steps in the optimization process for the purification of monoclonal antibodies are applicable to many other biomolecules. The main contaminants in preparations of monoclonal antibodies are albumin, transferrin and host immunoglobulins, all of which have different hydrophobicities. If the conditions are optimized so that protein binding capacity of the adsorbent is used mainly for binding the immunoglobulin, the productivity increases.

First, the selectivity of two hydrophobic ligands, alkyl and phenyl, was examined. As shown in *Figure 15*, both IgG and albumin eluted as narrower peaks on Alkyl Superose® than on Phenyl Superose®. Second, starting conditions were evaluated to determine the optimal concentration of ammonium sulfate used for equilibration of the Alkyl Superose column that would allow the immunoglobulin to bind while most of the contaminants pass through. *Figure 16* shows that in 2 M ammonium sulfate, both albumin and IgG were adsorbed and eluted in the decreasing salt gradient. In 1.5 M ammonium sulfate, the interaction was weaker, less albumin bound and the IgG eluted earlier in the salt gradient. In 1 M ammonium sulfate, albumin passed through the column, but the IgG still bound. In 0.8 M ammonium sulfate, IgG was retarded but eluted in a broad peak. (*Note*: The effectiveness of different salts in promoting and regulating hydrophobic interactions varies. For further information, see Ref. 17.)

Immunoglobulins are relatively hydrophobic and precipitate at fairly low salt concentrations, but will bind to hydrophobic adsorbents at even lower salt concentrations. To avoid precipitation of the monoclonal antibody, which can cause loss of specific activity, a lower salt concentration (0.8 M) was used for sample application than for column equilibration. After binding to the adsorbent, the monoclonals are more stable in the higher ammonium sulfate concentrations.

When sample is applied at a salt concentration lower than that used for the equilibration of the column, the sample volume becomes important (*Figure 17*). For example, when a 500 μl sample of ascites fluid was applied to a 1 ml column of Alkyl Superose, albumin, the weakest interacting substance, started to elute during sample application (*Figure 17a*). Dividing the sample into portions, e.g. five 100 μl samples, and adding equilibration buffer (1.3 ml in this case) after each sample application to enhance the hydrophobic interaction prevented early elution of sample (*Figure 17b*).

When the temperature changes, the separation achieved by hydrophobic interaction chromatography also changes. On a

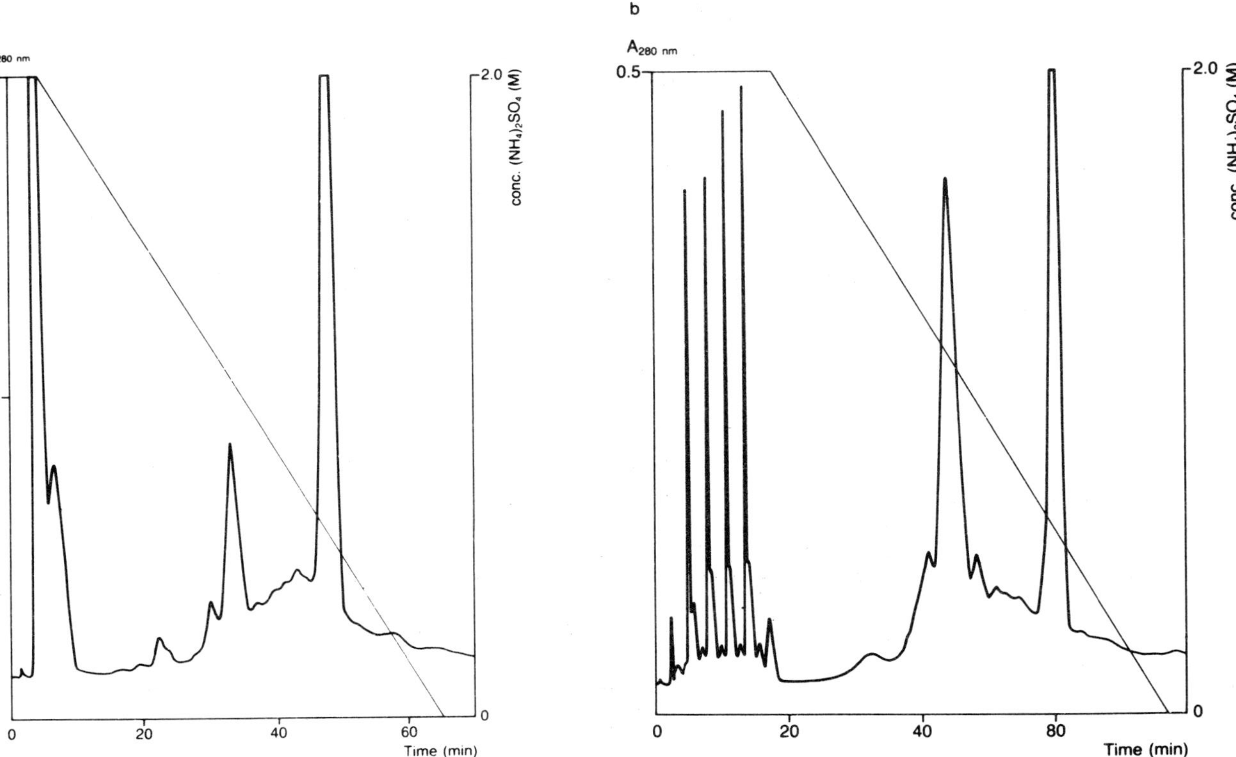

Figure 17. The effect of loading conditions in HIC. Column, Alkyl Superose HR 5/5; flow rate, 0.5 ml min^{-1}; buffer A, 0.1 M phosphate, pH 7.0, $(NH_4)_2SO_4$. (a) Sample (500 μl anti-CEA MAB (IgG_1) from mouse ascites fluid in 0.9 M $(NH_4)_2SO_4$ (corresponding to 115 μl ascites)) applied in one injection. (b) Sample (as (a)) applied in five 100 μl injections with 1.3 ml 2.0 M $(NH_4)_2SO_4$ after each portion.

laboratory scale, elution can be achieved by moving a packed column to which sample has been applied into a cold room. For large scale chromatography, this is impractical, and it is essential to keep the temperature constant during scale up.

LINKING OPTIMIZED STEPS

Once the steps are optimized, they should be put together in a logical fashion. Linking of the different steps in an efficient manner is crucial to achieving the objectives that are common to all chromatographic separation schemes (*Table 9*).

The initial process volume will usually be large. Removing the molecule of interest from most of the contaminants using highly selective techniques will reduce the process volume and concentrate the product. Affinity chromatography is the most selective technique and can be an excellent first step. For example, Protein A Sepharose® has been used for the large-scale preparation of highly purified monoclonal antibodies.[18] A bound receptor has been used for the one-step purification of recombinant interleukin-2.[19] Affinity media can be rather expensive, but their ability to reduce the number of chromatographic steps may justify the expense. It is advisable to use an affinity gel that has a stable ligand. Lectin affinity media, for example, may not have a long life, and proteinaceous ligands in general are subject to degradation by the presence of proteases found in the initial feed. Protein A is unusual because it is a very stable protein, but it, too, will be subject to the action of proteases. Dye ligands are fairly stable, and their low cost makes them particularly suitable for process chromatography. Due to the cost and life of most affinity media it is advisable to remove the bulk of contaminants first.

Ion exchange chromatography is an excellent first step because of its widespread applicability, high capacity, and relatively low cost. New rigid ion exchangers can be used in automated column techniques. In some cases, by selecting the appropriate anion or cation exchanger the bulk of the contaminants can be bound to the exchanger, while the molecules of interest pass through. The nonbinding materials can then be bound to an ion exchanger with the opposite charge to obtain further purification and concentration. Or the same type of ion exchanger can be used at a different pH

Table 9. Objectives in a chromatographic purification scheme.

Obtain the required degree of purity
Obtain sufficient quantity of purified material
Preserve biological activity
Produce the product in an efficient and cost-effective manner

or ionic strength so that the molecules of interest bind. To obtain the maximum capacity from an ion exchanger, use the highest ionic strength which still allows the product to bind but enables as many contaminants as possible to pass through. The product can then be eluted with a minor change in ionic strength and/or pH. Other contaminants will remain bound to the exchanger and can be eluted with high salt buffer. As a guideline, stay within one pH unit of the isoelectric point of the molecule of interest.

With the correct pH and ionic strength it is sometimes possible to use both anion and cation exchangers to bind the product. An example of this strategy is shown in *Figure 18* for the purification of human interleukin 1 (IL-1).[20] In the first step, the culture supernatant was applied to a sulfopropyl (SP) cation exchanger. The IL-1 bound while some contaminants were removed. A change of buffer and pH caused the IL-1 to elute. The buffer used for elution from the cation exchanger was the same buffer used to equilibrate the next column—a DEAE anion exchanger. The IL-1 was bound to the DEAE exchanger and then eluted in a sharp peak by increasing the ionic strength. After a reduction in ionic strength, the IL-1 was applied to a Procion Red agarose affinity column. In these three steps, IL-1 was purified to homogeneity, as determined by polyacrylamide gel electrophoresis.

In most cases, it is desirable to use a different mode of separation in each succeeding step. *Figure 19* shows an example of such a scheme developed for a laboratory scale separation of colony-stimulating factor.[21] After ammonium sulfate fractionation, hydrophobic interaction chromatography was used to bind the colony stimulating factor. This step was followed by ion exchange chromatography, and then by affinity chromatography. The purification scheme took advantage of hydrophobicity, charge and biospecificity of the protein. The final product was purified 200-fold.

Figure 18. Human interleukin 1: purification to homogeneity.

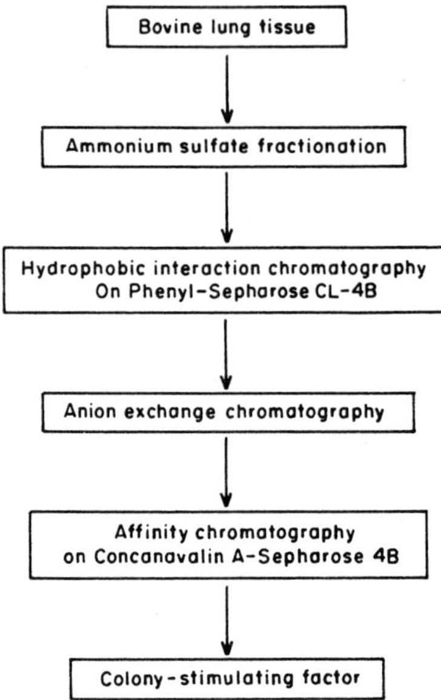

Figure 19. Isolation of a high molecular-mass, granular protein. (From Neumeir, R. and Maurer, H. R. *Hoppe-Seyler's Physiol. Chem.* **363** (1982) 1493–1500.)

The chromatography steps should be linked so that if possible the sample from one step is ready for application to the next step. This can eliminate the need for concentration and buffer exchange. For example, if hydrophobic interaction chromatography is followed by ion exchange chromatography, the sample is applied in high salt to the hydrophobic interaction medium and eluted in low salt. The low salt eluate can then be applied to the ion exchanger. (If the salt concentration is too high, the sample can easily be diluted.) Alternatively, if ion exchange chromatography is used first, the high salt eluate can be applied directly to a hydrophobic interaction column. If the salt concentration is not sufficient to promote binding of the sample, then salts can be easily added. This has been done in the purification of some isozymes.[22]

Some techniques lead to sample dilution. Dilution steps should be followed by concentrating techniques. Gel filtration used to fractionate molecules by size causes the sample to be diluted, and since the sample volume is limiting, this technique should be used only near the end of the purification scheme when the process volume is reduced. Fractionation by gel filtration should be followed by a step, such as ion exchange or affinity chromatography, that leads to concentration. While the molecule is being concentrated, it is also being removed selectively from its contaminants. Gel filtration is, however, often used as a final step to remove aggregates and transfer the product into its final packaging solution.

Table 10. Some logical sequences of chromatographic steps.

Sequence	Next step (alternatives)
Ion exchange → Hydrophobic interaction chromatography →	Chromatofocusing Affinity chromatography Gel filtration
Ammonium sulfate precipitation → Gel filtration → Ion exchange →	Affinity chromatography Hydrophobic interaction Chromatofocusing
Ammonium sulfate precipitation → Hydrophobic interaction chromatography →	Affinity chromatography Gel filtration Ion exchange
Ion exchange → Affinity chromatography	
Ion exchange → Affinity chromatography → Gel filtration	
Ion exchange → Gel filtration	

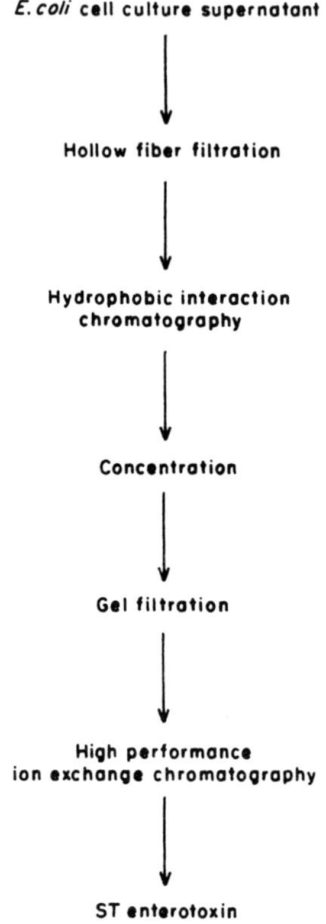

Figure 20. The purification of ST enterotoxin from *Escherichia coli*.

Gel filtration may also be used as a desalting or buffer exchange technique. In this manner, the flow rate is high and the capacity moderate. In addition, the dilution factor is not nearly as great. If the sample volume is 20–25%, the dilution factor is only 1.1–1.5. If it is not possible to directly link the chromatography steps, buffer exchange by gel filtration is a good tool for process chromatography. *Table 10* shows some possible logical sequences of chromatographic steps.

Generally, high performance chromatography steps should be used toward the end of the purification scheme as this technique can be expensive. Preliminary steps to reduce the bulk of contaminants and reduce the process volume can lengthen the life of high performance columns. A purification scheme which uses some of the principles discussed here is shown in *Figure 20.* The initial step in the purification of ST enterotoxin from *Escherichia coli* was centrifugation.[23] This was followed by sterile filtration and then hollow fiber ultrafiltration using a 10 000 nominal molecular weight cut-off membrane. The filtrate was adsorbed to a hydrophobic interaction medium. After elution from the column, the active fractions were concentrated and applied to a gel filtration column. After gel filtration, the sample was applied to a high performance anion exchange column. After the last step, which had a cycle time of only 20 min, the ST enterotoxin was judged pure by amino acid and UV spectral analyses.

For further reading, see Appendix A (An Overview of Chromatographic Techniques) and Ref. 24.

REFERENCES

1. Guiochon, G. and Colin, H. Theoretical concepts and optimization in preparative scale liquid chromatography. *Chromatogr. Forum* **1** (1986) 21-28.
2. Janson, J.-C. and Hedman, P. On the optimization of process chromatography of proteins. *Biotechnol. Prog.* **3** (1987) 9-13.
3. Note. While there have been some attempts to use computer simulation for optimization, in practice, optimization of each step can be accomplished in a very short time with minimal labor by using an automated chromatography system. By optimizing on a lab. scale, chromatographic media and buffer costs will be reduced, and valuable sample will be saved. Snyder, L. R. and Dolan, J. W. HPLC computer simulation: optimizing column conditions. *Am. Lab.* **18** (1986) 37-43. Evans, L. B. and Field, R. P. Bioprocess simulation: a new tool for process development. *Bio/technol.* **6** (1988) 200-203.
4. Chase, H. A. Prediction of the performance of preparative affinity chromatography. *J. Chromatogr.* **297** (1984) 179-202.
5. Arnold, F. H., Chalmers, J. J., Saunders, M. S., Croughan, M. S. *et al.* A rational approach to the scale-up of affinity chromatography. In *ACS Symposium Series*, No. 271 (LeRoith, D. Shiloach, J. and Leahy, T. J. eds). American Chemical Society, Washington, DC, 1985, p. 113.
6. Schmuck, M. N., Gooding, D. L. and Gooding, K. M. Comparison of porous silica packing materials for preparative ion-exchange chromato-graphy. *J. Chromatogr.* **359** (1986) 323-330.
7. Sitrin, R. D., DePhillips, P. A. and Dingerdissen, J. J. Preparative reversed-phase HPLC of polar fermentation products. *Develop. Ind. Microbiol.* **27** (1987) 65-75.
8. Johnson, R. D. Preparative high performance liquid chromatography of peptides and proteins. *Develop. Ind. Microbiol.* **27** (1987) 77-83.
9. Note. This is the maximum flow rate used for this experiment. It appears that even higher flow rates would give sufficient resolution.
10. Scale-up factors for ion exchange. *Separation News* **13.2**, March 1986.
11. Hagel, L. In *High Resolution Protein Purification* (Janson, J.-C. and Ryden, L. eds). Verlag Chemie, Deerfield Beach, in preparation.
12. Dolan, J. W., Snyder, L. R. and Quarry, M. A. HPLC method develop-ment and column reproducibility. *Am. Lab.* **18** (1987) 43-47.
13. Guo, D., Mant, C. T., Taneja, A. K. and Hodges, R. S. Prediction of peptide retention times in reversed-phase high performance liquid chromatography. *J. Chromatrogr.* **359** (1986) 519-532.
14. Meek, J. L. Prediction of retention times in high-pressure liquid chromatography on the basis of amino acid composition. *Proc. Natl Acad. Sci.* **77** (1980) 1632-1636.
15. Hoeger, C., Galyean, R., Boublik, J. M. *et al.* Preparative reversed phase high performance liquid chromatography: effects of buffer pH on the purification of synthetic peptides. *BioChromatogr.* **2** (1987) 134-142.

16. Steffensen, R. and Anderson, J. J., Jr. Reversed-phase separation of proteins. *BioChromatgr.* **2** (1987) 85-89.
17. *Octyl-Sepharose® CL-4B and Phenyl-Sepharose® CL-4B for Hydrophobic Chromatography.* Pharmacia, Uppsala.
18. Stephenson, J. R., Lee, J. M. and Wilton-Smith, P. D. Production and purification of murine monoclonal antibodies. *Anal. Biochem.* **142** (1984) 189-195.
19. Bailon, P., Weber, D., Keeney, R. F., Fredericks, J. E. *et al.* Receptor-affinity chromatography: a one-step purification method for recombinant interleukin-2. *Bio/technol.* **5** (1987) 1195-1198.
20. Kronheim, S. R., March, C. J., Erb, S. K., Conlon, P. J. *et al.* Human interleukin 1 purification to homogeneity. *J. Exp. Med.* **161** (1985) 490-502.
21. Neumeir, R. and Maurer, H. R. Isolation of a high molecular weight mass granulocyte colony-stimulating factor from bovine lung conditioned medium. *Hoppe-Seyler's Z. Physiol. Chem.* **363** (1982) 1493-1500.
22. Sullivan, D. M. and Hoffman, J. L. Fractionation and kinetic properties of rat liver and kidney methionine adenosyltransferase isozymes. *Biochem.* **22** (1983) 1636-1641.
23. Ronnberg, B. Improved method for purification of *Escherichia coli* heat stable enterotoxin by hydrophobic interaction, molecular-sieve, and high performance ion exchange chromatography. *Prep. Biochem.* **13** (1983) 245-260.
24. *Purification of Fermentation Products, ACS Symposium Series No. 271* (LeRoith, D., Shiloach, J. and Leahy, T. J., eds). American Chemical Society, Washington, DC, 1985.

7 Scale Up

Guidelines for scale up of chromatography are relatively simple. There are, however, some non-chromatographic factors that must also be considered to ensure that the entire process is scaled up efficiently with no loss of product activity. Although each production situation is unique, an idea of the scale, number of steps and process time is presented for production of some products of modern biotechnology. Scenarios are given for production of 100 g, 10 kg, and 500 kg of product per year.

GUIDELINES

After the purification scheme is optimized on a laboratory scale, the first scale up is usually on the order of 100-fold. Depending upon the quantity of product required and recycling capabilities, this may, in fact, be the final scale. For example, to scale up 100-fold, relevant

Table 11. Scale up guidelines.

MAINTAIN:
- Bed height
- Linear flow
- Sample concentration
- Gradient volume:media volume

INCREASE:
- Sample load
- Volumetric flow rate
- Column diameter

CHECK SYSTEM FACTORS:
- Distribution system
- Wall effects
- Piping (linear flow in whole system)

parameters such as sample load, volumetric flow rate, and gel volume will be increased 100-fold. The column bed height, linear flow rate, sample concentration and ratio of sample to gel—all optimized on a laboratory scale—will be kept the same (see *Table 11*). If a gradient is used for elution, the ratio of gradient volume to gel volume will remain constant and, therefore, the time required for the gradient to develop will remain about the same on the larger column.[1] Further scale up, from pilot plant to production, is usually 10- to 30-fold to ensure reproducibility.

In practice, an increase in column diameter will lead to a decrease in column wall support for the media. For example, at a given pressure, the linear flow rate usually falls 30–45% when the diameter is increased from 2.6 to 10 cm (see *Figure 21*). For diameters above 10 cm, the linear flow rate is almost independent of column width, particularly if the effects of pressure drop in valves and connections of the chromatographic system are minimized. However, if non-rigid

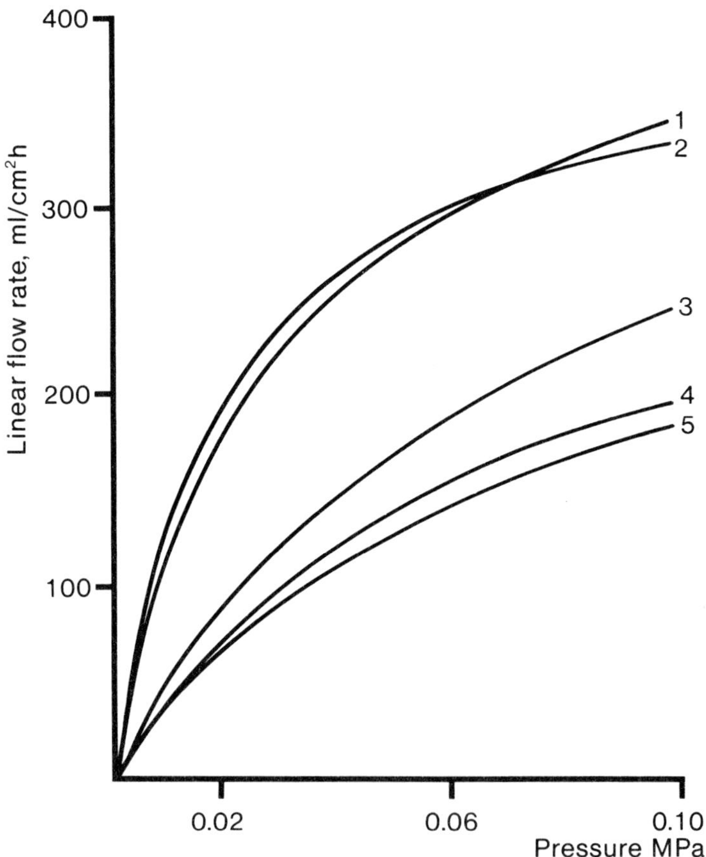

Figure 21. Linear flow rate at different column diameters and pressure drops. 15 cm packed bed of Sepharose 6 Fast Flow. Column diameters: curve 1, 5.0 cm; curve 2, 11.3 cm; curve 3, 25.2 cm; curve 4, 37.0 cm; curve 5, 80.0 cm.

chromatographic supports are used, the linear flow rate may have to be decreased from the optimal rate to minimize media compression. Sepharose® Fast Flow media were developed to allow direct scale up.

When scaling up, you must make sure that the larger equipment does not cause extra-column zone spreading, which leads to a loss in resolution. For example, changes in the size of the monitoring cell and different lengths and diameters of outlet pipes or tubings can cause zone spreading in the larger system.

If a deviation from the laboratory scale results are seen in the larger column, there are several factors to check. The monitoring system should be examined to verify that transport distances between column outlets, monitors and fraction collection valves do not introduce time delays or volume changes which make the process controller switch the fraction collector valve position at the wrong time. While this may not be detrimental to the actual operation of the column, it could result in false measurements of column efficiency (see Appendix D, Column Packing).

If the larger column has a less efficient flow distribution system than the analytical column, greater axial dispersion in the bed, as well as extra zone spreading in the end pieces, will occur. The extra zone spreading will always dilute the product fraction. Unless this dilution is acceptable, it is necessary to compensate by reducing one of the factors contributing to the total product peak variance. This could be achieved by reducing the flow rate (lesser contribution from intra-particle mass transfer resistances) or by reducing the sample size (lesser overload contribution). These changes will, of course, reduce the productivity.

If the zone spreading in gel filtration also increases the amount of contaminants in the product, then it is justifiable to increase the column length in order to increase the effective plate number. If all other parameters are held constant, this will result in an increased backpressure. (If, on the other hand, higher resolution than required is obtained, it is possible to decrease the column length to achieve greater throughput.)

The simple solution is to minimize broadening at larger scale by selecting optimal equipment and using the optimal system configuration. In fact, quite impressive results have been obtained using process columns. Sephamatic® stainless steel columns with a diameter of 100 cm have been packed with Sephacryl® S-200SF and HETP values of 0.034 obtained. These results correspond well with those achieved in laboratory size columns. *Figure 22* shows the comparable HETP values obtained with a sodium chloride sample in a 32 l column and in a 236 l column.

There are some unavoidable differences between analytical and production columns. Since the surface area per unit volume decreases with increasing size, there should be fewer problems in larger columns with nonspecific adsorption to column walls, inlet and outlet

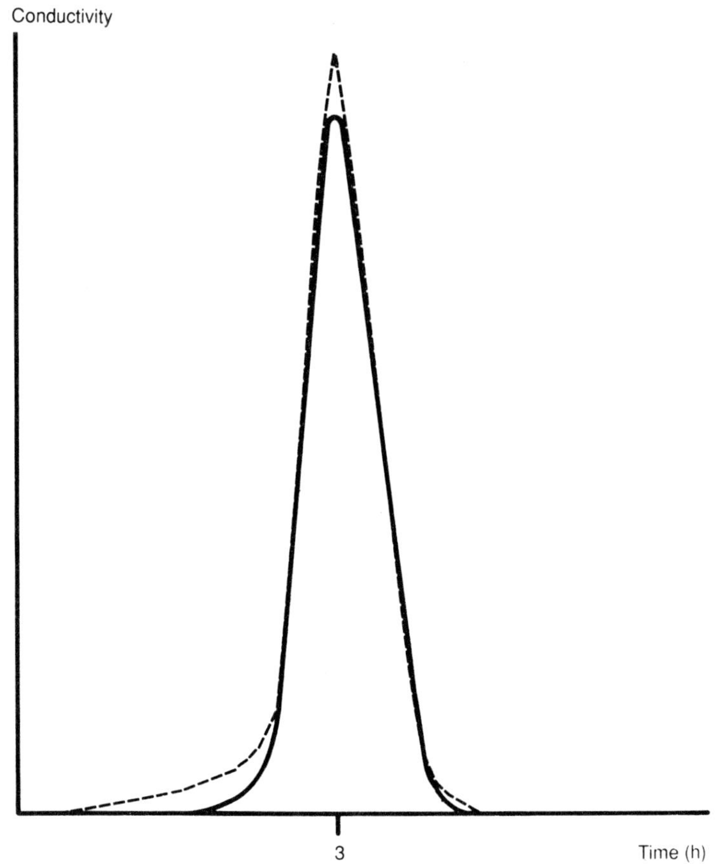

Figure 22. Comparison of HETP values in a 32 l column (—) and in a 236 l column (----), both with 30 cm bed heights. Sample, 2.5% V_t, 10 g l^{-1} NaCl; flow rate, 10 cm h^{-1}; medium, Sephacryl® S-200 Superfine; HETP, (—) 0.030 cm, (----) 0.034 cm; time, 3 h.

flow distributors, tubing or pipes, and other inline equipment. In the larger columns, it should also be possible to achieve optimal packing structure in a larger fraction of the bed since there will always be a larger void fraction close to the wall than in the center of the column.

At the laboratory scale, chromatographic steps are easily and rapidly optimized using high resolution (e.g. 10 μ) media. If the particle size is increased to achieve better throughput and reduced operating costs at the pilot stage, further optimization will be required. For media that are designed to make this transition, established guidelines may be used. As shown in *Figure 23* scale up from Mono Q® to Q Sepharose® Fast Flow is easily achieved by following established guidelines for scale up.

An alternative, especially if media series such as Mono Q and Q Sepharose Fast Flow are not available, is to optimize at a laboratory scale with the media that are to be used in the final process.

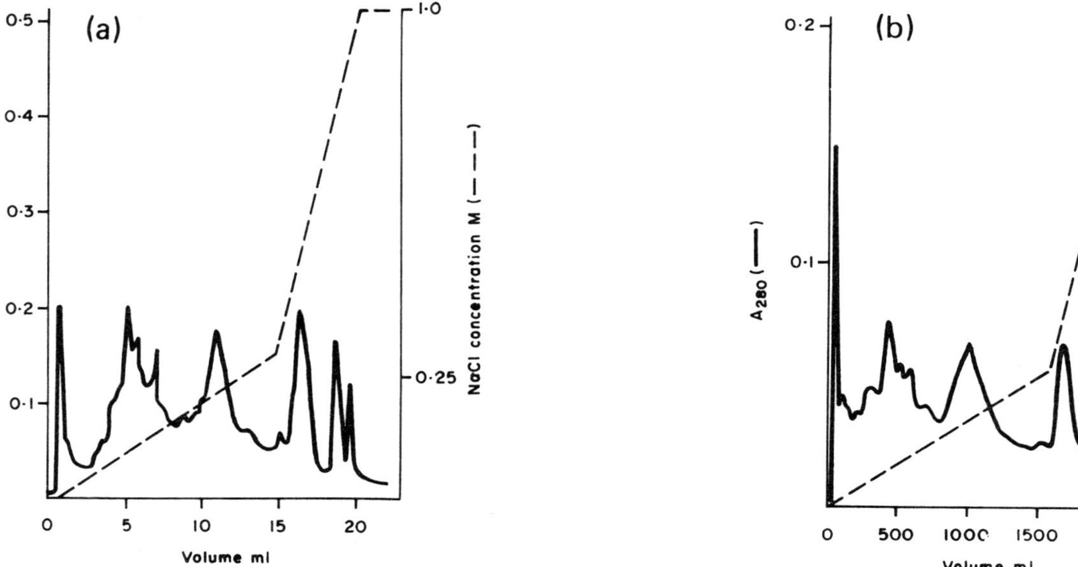

Figure 23. Scale up from Mono Q® to Q Sepharose® Fast Flow. (a) Medium, Mono Q®; sample, bovine liver extract (1.6 mg protein); sample volume, 100 μl; column, HR 5/5; bed dimensions, 0.5 × 5 cm; buffer A, 50 mM diethanolamine, pH 8.8; buffer B, A + 1.0 M NaCl; gradient volume, 20 ml; flow rate, 1.0 ml min^{-1}; time, 21 min. (b) Medium, Q Sepharose® Fast Flow; sample, bovine liver extract (83 mg protein); sample volume, 5.3 ml; column, K 26/40; bed dimensions, 2.6 × 10 cm; buffer A, 50 mM diethanolamine, pH 8.8; buffer B, A + 1.0 M NaCl; gradient volume, 2200 ml; flow rate, 5.3 ml min^{-1}; time: 420 min.

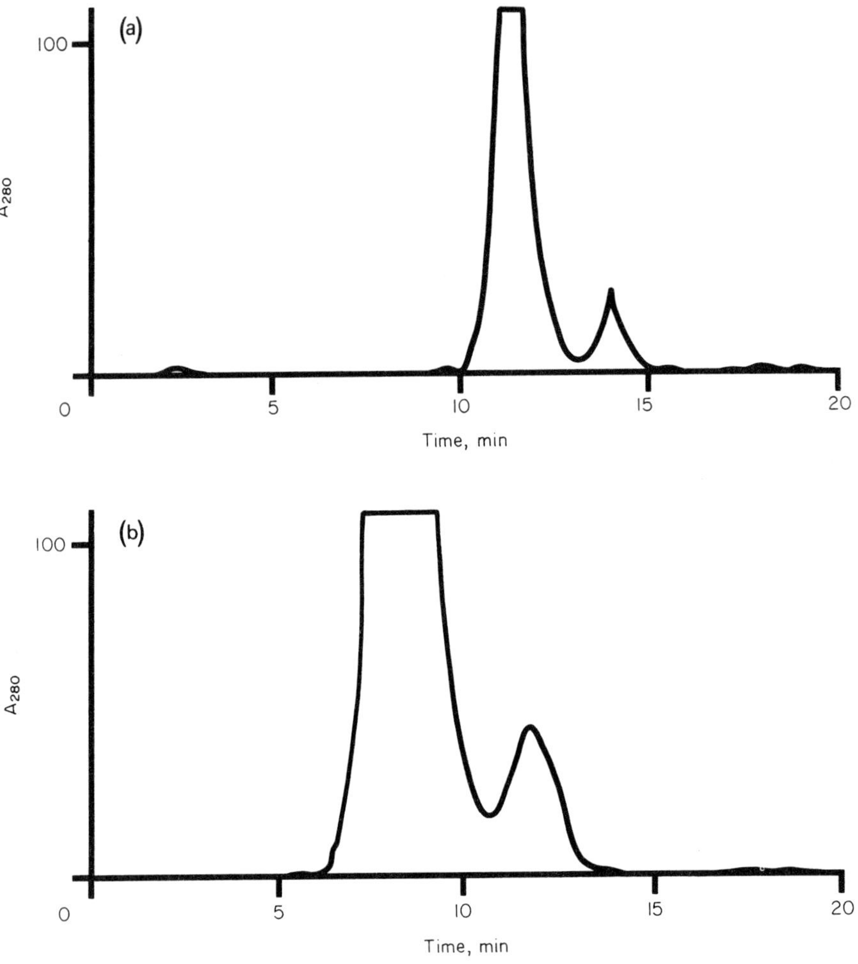

Figure 24. Separation of monoclonal antibody from free light-chain by high performance gel filtration. (a) Column, Superose® 6 HR 10/30; flow rate, 1 ml min^{-1}; sample volume, 100 μl (200 mg ml^{-1}). (b) Column, Superose® 6 prep. grade HR 16/50; flow rate, 1 ml min^{-1}; sample volume, 5 ml (2 mg ml^{-1}).

At the pilot scale, some refinements of the purification scheme may be made. It is interesting to note (*Figure 24*) the scale up of a high performance gel filtration separation of a monoclonal antibody from free light-chain. In *Figure 24a*, a 10 μ bead is used and the sample load is 0.4% of the total bed volume; in *Figure 24b*, a 30 μ bead is used and the sample load is increased to 5% of the total bed volume. The matrices are essentially identical except for the bead diameters. From this example it can be seen why it is important to do some further optimization at the scale-up stage if the media are changed. In addition, any changes in the chromatographic materials, such as ligand density, during scale up will certainly have a profound effect on the results.

The scale up of chromatographic processes has, in fact, been shown to be straightforward. Non-chromatographic factors can, however, alter the chromatography during scale up. These factors include: changes in sample composition and concentration that often occur as the fermentation scale increases, precipitation in the biological feedstock due to longer holding times when large volumes must be handled, non-reproducibility of the buffer quality due to inadequate facility to prepare large quantities of buffer consistently, and microbial growth in the buffer due to increased handling and longer holding times.[2] Additives may change buffer pK_a values.[3] Finally, changes in temperature, pH or ionic strength may alter the sample and, as a result, the chromatography.

When the guidelines given in *Table 11* are followed and equipment carefully selected, it is possible to maintain the same degree of resolution at the larger scale. However, very often (especially in larger companies) the personnel responsible for the pilot plant or production stages are not the personnel responsible for developing the purification scheme at the laboratory scale. The goals of the pilot plant engineer are often different from those of the researcher who has developed the purification scheme. Whereas the researcher may strive for the resolution of the maximum number of components, at the pilot plant or production level it is necessary to separate the component of interest from the bulk of impurities with the maximum sample load possible. It is, therefore, important to encourage communication between the two groups so that the goals are mutual.

The process will be verified at the pilot plant stage, and automation (see Chapter 8) and quality control should be part of the process. In addition, it is at this scale that material for clinical trials will usually be produced, thus there is often a need for increased documentation and for establishing and maintaining sanitary conditions at the pilot plant scale (see Chapter 9).

Figure 25 shows the 250-fold scale up of ion exchange chromatography used in the chromatographic fractionation of desalted plasma. In this step, IgG, albumin and glycoprotein are separated. The bed height (15 cm) and linear flow rate (120 cm h^{-1}) were maintained constant as the process was scaled up from columns of 30 ml to 1.5 l to 7.5 l. As can be seen from the chromatograms, resolution was maintained at each step.

To ensure that the scaled up process is reliable and reproducible, continuous monitoring of flow rate, pressure, tank levels, presence of air in the system, feedstock concentration, conductivity and pH may be required (see Chapter 8). Proper column packing and cleaning of the media in the column, as well as the entire system, will also affect the reproducibility of the process (see Chapter 9). Chromatographic media must be capable of being cleaned in place (CIP), unless a batch mode is used. Media that are produced according to good manufacturing practice (GMP), and quality controlled for relevant parameters, are required for the reproducible production of biologicals. Proper documentation should be provided by the vendor. In the production of a therapeutic, regulatory agencies often require documentation that shows the absence of leakage products from the chromatographic media in the final product. It may be necessary to perform extractables tests (US Pharmacopoeia). Toxicity data on chromatographic media may also be required, especially if a potentially toxic ligand is used.

PRODUCTION SCENARIOS

The following scenarios present guidelines for the production of three types of products of modern biotechnology: a mammalian cell culture system in which production requirements are 100 g year^{-1}, a prokaryotic system producing 10 kg year^{-1} and a prokaryotic system producing 500 kg year^{-1}. These guidelines are presented to give a sense of scale of operation and do not take into account biological activity and final product quality.

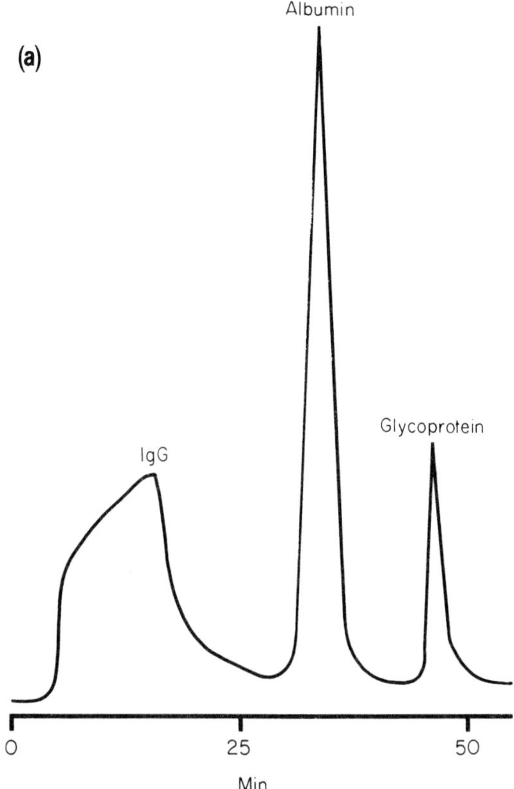

Figure 25 (See page 63 for caption)

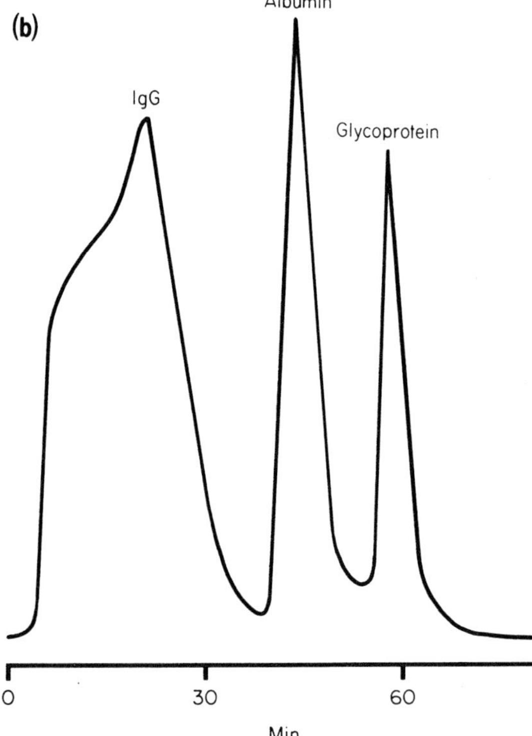

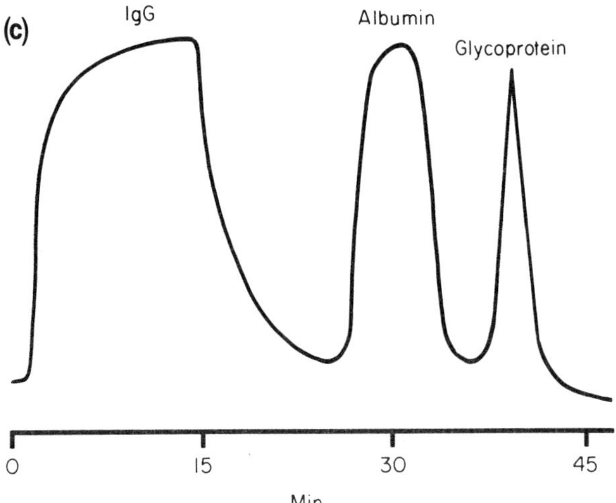

Figure 25. Scale up of ion exchange chromatography. (Work from Pharmacia LKB Biotechnology AB.) Sample, desalted plasma; medium, DEAE Sepharose® Fast Flow; flow rate, 120 ml cm^{-2} h^{-1}; bed height, 15 cm. (a) Column, K 16 (16 mm diameter); loading, 1.2 g protein. (b) Column, BP 113 (diameter 113 mm); loading, 90 g protein. (c) Column, BP 252 (diameter 252 mm); loading, 330 g protein.

Table 12. Large scale purification scenario: mammalian expression system. Product volume approximately 100 g y^{-1}.

	Concentration	Desalting/ conditioning	Initial purification: ion exchange	Final purification: gel filtration
Input	100 l 12 g protein 3 g product	1 l 11.5 g protein 2.9 g product	2 l 11.3 g protein 2.9 g product	0.6 l 3.5 g protein 2.8 g product
Output	1 l 11.5 g protein 2.9 g product	2 l 11.3 g protein 2.9 g product	0.6 l 3.5 g protein 2.8 g product	1.5 l 2.6 g protein 2.6 g product
Total process time	**—**	**2.5 h**	**1.5 h**	**12 h**
Cycles	**—**	**5 cycles**	**1 cycle**	**2 cycles**
Flow rate	**—**	**18 l h^{-1}**	**6 l h^{-1}**	**1 l h^{-1}**
Column volume	**—**	**6 l**	**300 ml**	**12 l**

Table 13. Large scale purification scenario: prokaryotic expression system. Production volume approximately 10 kg y^{-1}.

	Enrichment: ion exchange	Fractionation: ion exchange	Concentration	Final purification: gel filtration
Input	250 l 1000 g protein 200 g product	40 l 400 g protein 180 g product	15 l 200 g protein 160 g product	3 l 190 g protein 155 g product
Output	40 l 400 g protein 180 g product	15 l 200 g protein 160 g product	3 l 190 g protein 155 g product	9 l 149 g protein 140 g product
Total process time	**2 h**	**2 h**	**2 h**	**10 h**
Cycles	**1 cycle**	**1 cycle**	**—**	**—**
Flow rate	**600 l h^{-1}**	**180 l h^{-1}**	**—**	**7.5 l h^{-1}**
Column volume	**30 l**	**15 l**	**—**	**75 l**

Production quantities of 100 g y^{-1} are estimated to be needed for products such as monoclonal antibodies. In a typical mammalian expression system, such as Vero, CHO or hybridoma cells, the product concentration could be on the order of 0.03 g l^{-1}, with total protein concentration of 0.12 g l^{-1}. (This assumes that the product is 25% of total protein.) One-hundred liters of cell culture supernatant could be processed per week for 40 working weeks (taking into account time for cleaning and repair of the plant). The process includes: clarification, concentration, desalting/conditioning, initial purification (ion exchange chromatography) and final purification by gel filtration. The inputs and outputs are given in *Table 12*.

Expression systems such as *Escherichia coli, Saccharomyces cerevisiae*, or *Aspergillus nidulans* might be used to produce substances such as human growth hormone or enzymes in quantities of about 10 kg y^{-1}. If the product is expressed at about 20% of total protein, then with a fermenter volume of 1250 l run once a week, 250 l of clarified cell lysate with product concentration of 0.2 g l^{-1} and total protein of 1 g l^{-1} could be produced. The steps used in this process are: harvesting of cells and clarification; enrichment; fractionation; concentration; final purification. The following 5-day working week schedule could be used to achieve the production goals.

Day 1: harvest
Day 2: disintegration, clarification and enrichment
Day 3: fractionation, concentration and final purification
Day 4: completion of final purification; cleaning, maintenance and quality control
Day 5: sterilization

Table 14. Large scale purification scenario: prokaryotic expression system. Production volume approximately 500 kg y^{-1}.

	Enrichment/ concentration: ion exchange	Initial purification: ion exchange	Intermediate purification: ion exchange	Final purification
Input	16 000 l 90 kg protein 17.9 kg product	1500 l 50 kg protein 15 kg product	800 l 15 kg protein 13 kg product	200 l 12 kg protein 11.5 kg product
Output	1500 l 50 kg protein 15 kg product	800 l 15 kg protein 13 kg product	200 l 12 kg protein 11.5 kg product	1000 l 11.5 kg protein 11.5 kg product
Total process time	8 h	8 h	6 h	24 h
Cycles	8 cycles	8 cycles	4 cycles	4 cycles
Flow rate	3000 l h^{-1}	1800 l h^{-1}	400 l h^{-1}	50 l h^{-1}
Column volume	200 l	100 l	50 l	1250 l

The estimated inputs and outputs, number of cycles, column volumes, and flow rates are given in *Table 13*.

For the production of large quantities (500 kg y^{-1}) of substances such as porcine or bovine growth hormone at an expression level of about 20% of total protein in an organism such as *E. coli* or *A. nidulans*, a fermenter volume of approximately 90 000 l (90 m^3) per week would be required. This would give a clarified broth volume of 16 000 l (16 m^3) in which the total protein content is 90 kg, of which 17.9 kg is the desired product. In addition to harvesting and clarification, the following steps might be used: enrichment and concentration by ion exchange chromatography, initial purification by ion exchange and a final purification by gel filtration. (For details, see *Table 14*.)

By following scale up guidelines and being aware of pitfalls, chromatography is readily scaled up for safe production of biologicals. Today, large scale chromatography is used for the production of vaccines, monoclonal antibodies, immune modulators and hormones. For further reading, see Ref. 4.

REFERENCES

1. Schmuck, M. N., Gooding, K. M. and Gooding, D. L. Preparative liquid chromatography of proteins. *J. Liquid Chromatogr.* **7** (1984) 2863-2873.
2. Johansson, H., Lindquist, L.-O. and Low, D. Design criteria for the purification of biopolymers for the pharmaceutical industry. Eurochem—Process Engineering Today, Birmingham, England, 3-5 June 1986.
3. Rubino, J. T. The effects of cosolvents on the action of pharmaceutical buffers. *J. Parenteral Sci. Technol.* **41** (1987) 45-49.
4. *Large-scale Adsorption and Chromatography* (Wankat, P. C., ed.). CRC Press, Boca Raton, 1986, Vols I and II.

8 *Equipment*

This chapter describes chromatographic equipment requirements for laboratory scale process development and optimization, pilot plant scale up and production. Also presented are guidelines for selecting pilot and production systems and components, and the advantages and disadvantages of several types of automation approaches are discussed. Validation, which is essential for the production of a pharmaceutical, is also described.

EQUIPMENT REQUIREMENTS AT DIFFERENT SCALES

The need for a flexible system is great during the initial development stage. System flexibility requires the ability to work with different chemicals, different particle sizes and different column sizes. The system should also be easily reconfigured to allow for optimizing all the steps in the purification process. A highly automated *laboratory* system provides maximum flexibility.

Good documentation of all parameters is required for further development of a process. The laboratory scale process development system should allow automatic documentation of volumes, monitor signals, and flow rates. Automation software must also be flexible and 'user-friendly' to simplify the search for optimal chromatographic media and purification conditions. An automated system can definitely minimize process development and optimization time.

During *pilot plant* scale up of an optimized purification protocol, the need for flexibility is generally reduced. However, when pilot plant purification systems are used for multiple purposes, flexibility remains an important factor. If the pilot plant equipment is also used for clinical trials production and other small scale production, the needs for documentation, reproducibility and validation become as important as in production scale systems.

At *production* scale, the need for flexibility decreases, but the need for reliability increases. Specifications must be set, and the equipment must be capable of providing the desired production capacity. For the production of a parenteral product, the equipment must also be capable of producing product at a specified level of purity in a safe and hygienic way. Equipment must be designed to prevent the escape of harmful materials from the system or their entry into the system.[1] Effective cleaning and, if necessary, sterilization must be possible, and biological activity and resolution must be preserved. Routine adjustments, calibration and preventative maintenance become essential.[2]

Regardless of scale, the separation power of a complete chromatographic system is determined by the chemical interactions between separation media and the molecules to be separated. The equipment does not contribute to the actual separation. Still, the same separation chemistry can give different performance when run in different equipment. This performance difference is a result of zone spreading created by backmixing effects in the liquid handling system or in the column inlets and outlets.

Zone spreading at different scales can be investigated by running

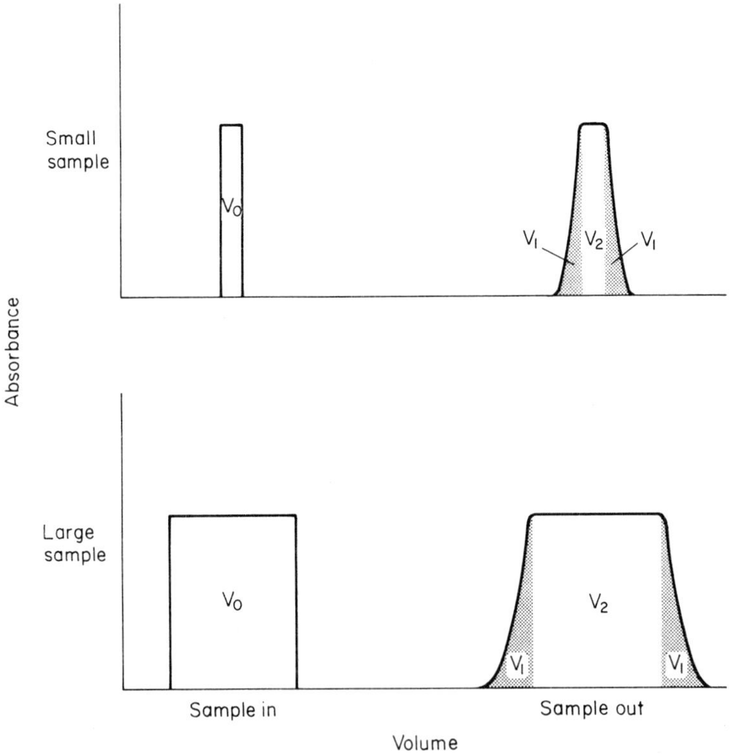

Figure 26. The effect of sample volume on zone spreading: V_0, sample volume; V_2, V_0 superimposed on 'sample out'; V_1, zone spreading (or peak broadening) on 'sample out'.

salt pulses through the system without a column. Zone spreading of the liquid handling equipment is measured as the dilution factor, which is equal to the volume of the sample leaving the system divided by the volume of the sample entering the system. To obtain a relevant measurement, the incoming sample volume should be the same as the volume of the actual sample to be separated in the system.

Zone spreading in analytical chromatography has been thoroughly investigated.[3] In general, extra-column zone spreading causes fewer problems in large-scale pilot and production chromatography. Dilution from extra-column zone spreading has less effect on larger peak volumes (*Figure 26.*)

In adsorption techniques, equipment installed upstream of the column has minimal dilution effects on the separations performed in the column. This is due to the concentrating effect the actual adsorption has on the sample. Major equipment-related zone spreading is due to dilution in the column outlet, the monitoring system, or the fraction collecting system. To minimize these effects, monitor flow cells, tubing and fractionating valves should be designed to minimize internal volumes.

In gel filtration and other partitioning techniques, the dilution effects of the whole equipment system (from the sample tank to the fractionating valves) can effect zone spreading. Therefore, partitioning techniques are generally more affected by the equipment during scale up, and greater care should be taken in the system design.

GUIDELINES FOR SELECTING PILOT PLANT AND PRODUCTION CHROMATOGRAPHY EQUIPMENT

A thorough determination of specifications for pilot plant and production systems can lead to the purchase of the most effective and cost efficient equipment. Functional, chemical and pressure specifications as well as hygienic requirements should be clearly examined.

Functional specifications

When choosing equipment for a pilot plant or a production plant, structure the system by functional units, e.g. for liquid delivery, separation, monitoring, fraction collection and control. *Figure 27* shows a typical functional system layout. For each unit, define in detail what tasks need to be performed as a 'functional specification'. This should lead to a detailed equipment and specifications list. For example, the liquid delivery unit can contain: buffer tank outlet valves and low-level alarms, sample tank drainage alarm, pump with flow rate control, in-line filters for buffers and sample, air bubble alarm, air trap, over-pressure alarm, filter by-pass, automatic high pressure filter switch, manual valve override, manual drainage valves, etc.

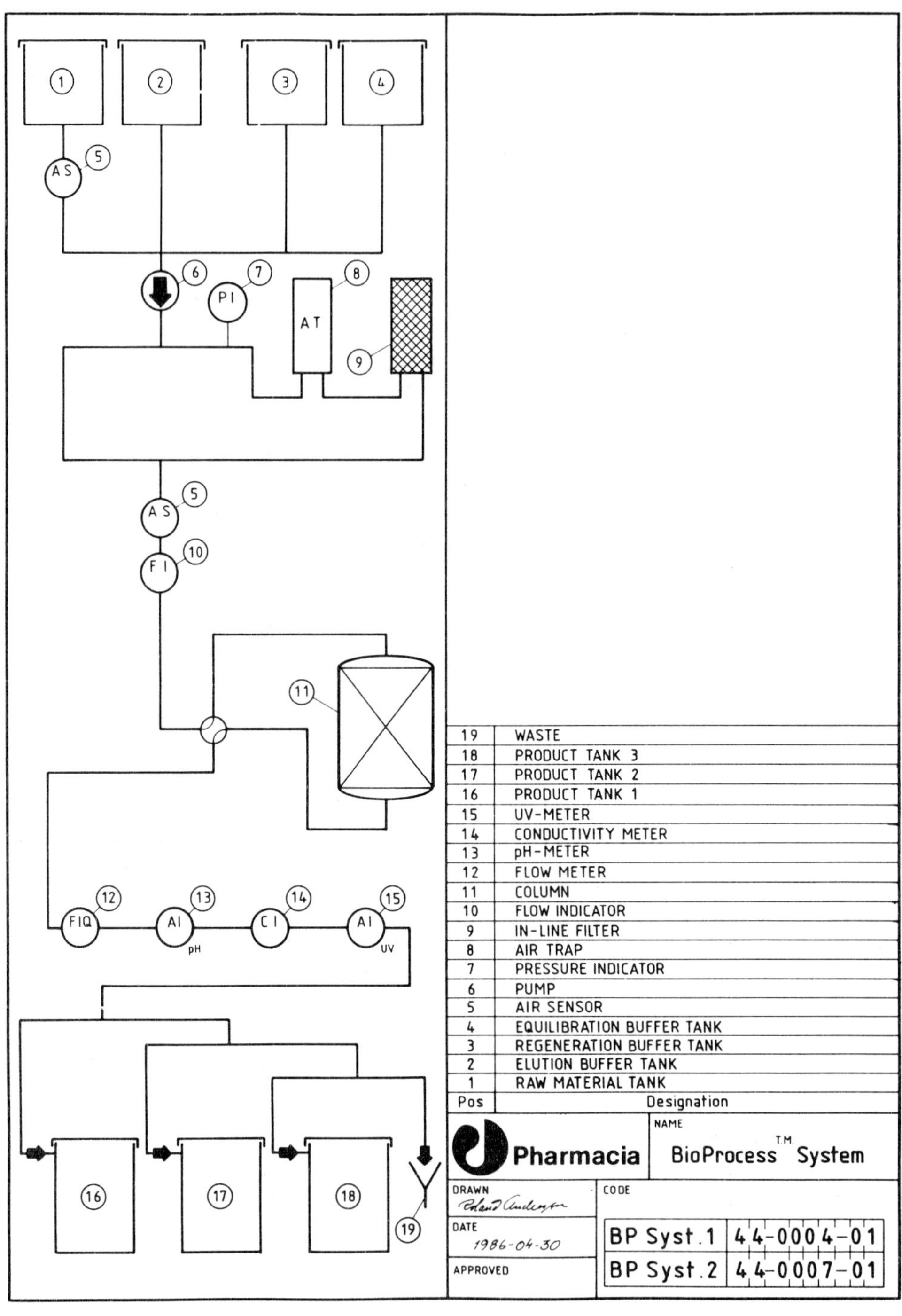

Figure 27. A standard flow sheet for a process chromatography system.

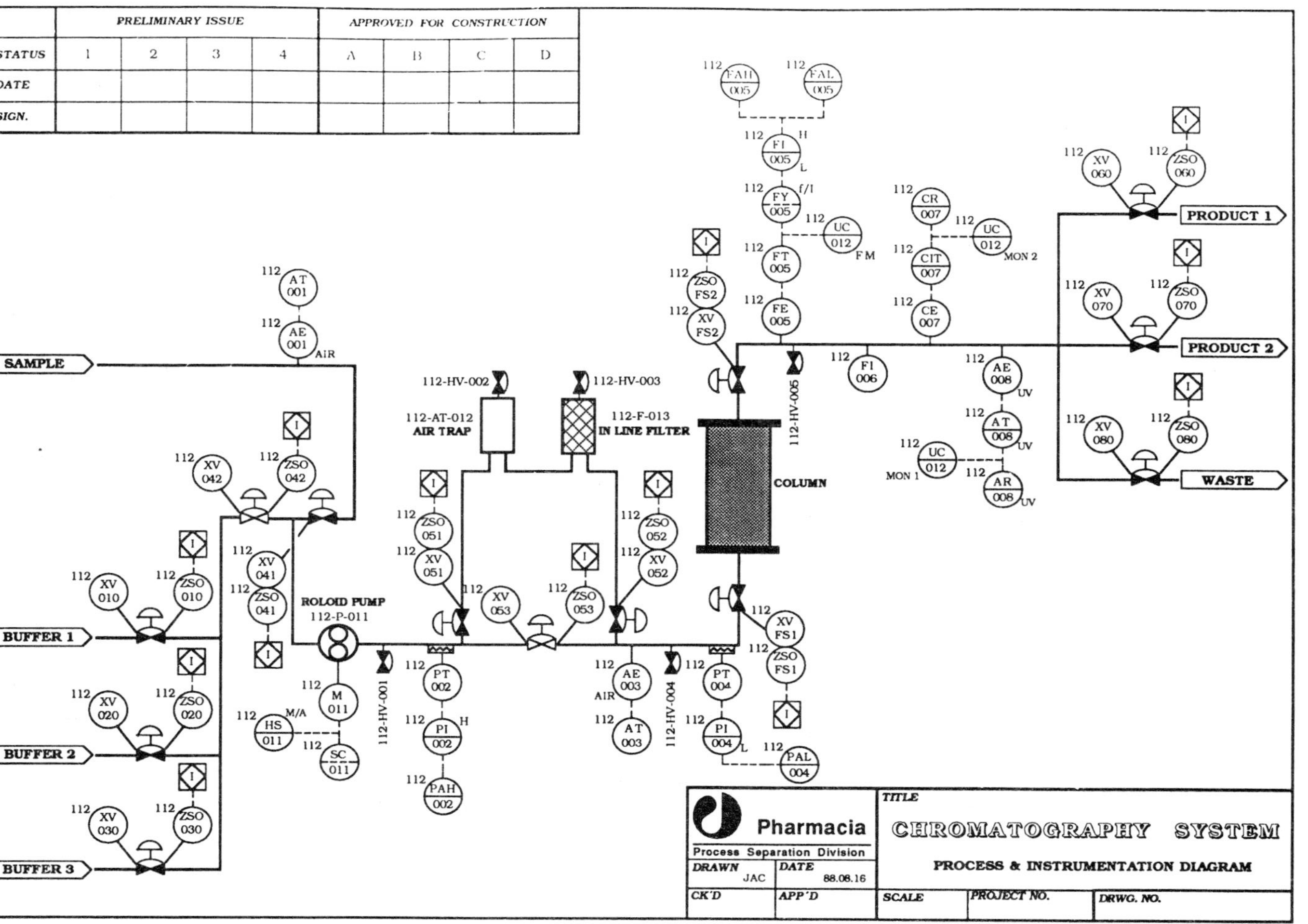

Figure 28. (See page 73 for caption)

GENERAL SYMBOLS			
1 PROCESS LINE	21 AUXILIARY OR UTILITY LINE	41 JACKETED LINE	61 STEAM TRACED LINE
2 ELECTRICALLY TRACED LINE	22 FUTURE LINE	42 VENDOR PACKAGE LIMIT	62 INSULATED LINE
3 GATE VALVE GENERAL SYMBOL	23 NO NORMALLY OPEN	43 NC NORMALLY CLOSED	63 DIAPHRAGM VALVE
4 BALL VALVE	24 BUTTERFLY VALVE	44 NEEDLE VALVE	64 CHECK VALVE
5 ANGLE VALVE	25 3-WAY VALVE	45 4-WAY VALVE (ALT. 1)	65 4-WAY VALVE (ALT. 2)
6 FLUSH BOTTOM TANK VALVE	26 VACUUM RELIEF VALVE	46 PRESSURE RELIEF OR SAFETY VALVE	66 S STEAM TRAP
7 IN LINE FILTER	27 A AIR TRAP	47 SG SIGHT GLASS	67 BELLOWS EXPANSION JOINT
8 HOSE	28 HOSE CONNECTION MALE	48 HOSE CONN. FEMALE	68 SLOPE % SLOPE (INDICATE %)
9 VENT	29 PULSATION DUMPER	49 Y-type STRAINER	69 T-type STRAINER
10 EXCHANGER UNIVERSAL	30 EXCHANGER SHELL/TUBE	50 EXCHANGER PLATE	70 F FLAME ARRESTOR
11 CONCENTRIC REDUCER	31 ECCENTRIC REDUCER	51 D CAP	71 FLANGE
12 CENTRIFUGAL PUMP	32 RECIPROCATING PUMP	52 ROTARY PUMP	72 PUMP UNIVERSAL SYMBOL
13 COMPRESSOR UNIVERSAL SYMBOL	33 COMP COMPRESSOR CENTRIF.	53 COMP COMPRESSOR ROTARY	73 COMP COMPRESSOR RECIP.
14 SERV. DESIGNATION DWG. NO. PROCESS TIE-IN	34 SP PIPING SPECIAL ITEM	54 CA SD DENOTES PIPING SPEC CHANGE	74 A REVISION SYMBOL
15 RUPTURE DISK PRESSURE	35 RUPTURE DISK VACUUM	55 MOTOR	75 CHEMICAL SEAL
16 PS PROCESS SEWER	36 CS CONTAINMENT SEWER	56 TANK INSULATION	76 AGITATOR
17	37	57	77
18	38	58	78
19 TANK CONE BOTTOM	39 TANK CLOSED	59 PRESSURE VESSEL	79 PRESSURE VESSEL JACKETED
20 CENTRIFUGE DISK	40 COLUMN	60	80

INSTRUMENT SYMBOLS		
81 CONNECTION TO PROCESS	103 CAPILLARY TUBING	125 SOFTWARE CONNECTION
82 ELECTRIC SIGNAL	104 ELECTROMAGNETIC OR SONIC SIGNAL	126 PNEUMATIC SIGNAL
83 INSTRUMENT LOCALLY MOUNTED	105 CENTRAL INSTRUMENT PANEL MOUNTED	127 CENTRAL INSTRUMENT BEHIND PANEL
84 INSTRUMENT LOCAL PANEL MOUNTED	106 INSTRUMENT BEHIND LOCAL PANEL	128 COMPUTER I/O INTERFACE
85 I PROGRAMMABLE CONTROLLER INTERLOCK	107 OPERATOR STATION POINT	129 LIGHT
86 ACTUATOR CYLINDER	108 ACTUATOR CYLINDER SINGLE-ACTING	130 ACTUATOR CYLINDER DOUBBLE ACTING
87 M ACTUATOR MOTOR	109 S ACTUATOR SOLENOID	131 ACTUATOR HAND
88 ACTUATOR DIAPHRAGM PRESSURE-BALANCED	110 ACTUATOR DIAPHRAGM	132 E/H ACTUATOR EL. HYDRAULIC
89	111	133
90	112	134
91	113	135
92	114	136
93	115	137
94	116	138
95	117	139
96	118	140
97	119	141
98	120	142
99	121	143
100	122	144
101	123	145
102	124	146

Figure 28. (See page 73 for caption)

The functional complexity determines the best automation structure for the system. Automation requirements are very different at different stages (see section on automation, page 86). For example, it is not usually necessary to automate alarms on buffer reservoirs for laboratory systems. On the other hand, a 1- or 2-day production loss may be incurred if a 1000 l column runs dry because the buffer tank is empty. As a general rule, functional complexity increases when going from laboratory scale to production, where reliability of the system is critical.

PIPING SERVICE DESIGNATION	
ACL	-CLEAN AIR
CIP	-CIP FLUID SUPPLY
CIPR	-CIP FLUID RETURN
CL	-CONDENSATE LOW PRESSURE
CH	-CONDENSATE HIGH PRESSURE
CO_2	-CARBON DIOXIDE
CS	-CONTAINMENT SEWER
CV	-CONTAINMENT VENT
CW	-CITY WATER
DIW	-DEIONIZED WATER
ET	-ETHANOL
FW	-FIRE PROTECTION WATER
HWS	-HOT WATER SUPPLY
HWR	-HOT WATER RETURN
IA	-INSTRUMENT AIR
N_2	-NITROGEN
P	-PROCESS
PA	-PLANT AIR
PS	-PROCESS SEWER WASTE
PV	-PROCESS SEWER VENTS
PW	-PROCESS WATER
SCL	-CLEAN STEAM LOW PRESSURE
SCH	-CLEAN STEAM HIGH PRESSURE
CC	-CLEAN CONDENSATE
SL	-STEAM 60 PSIG
SH	-STEAM 150 PSIG
ROW	-REVERSE OSMOSIS WATER
SW	-SANITARY SEWER
SV	-SANITARY VENT
ST	-STORM VATER
VT	-VENT
WP	-PORTABLE WATER
WPH	-PORTABLE WATER HOT
WFI	-WFI WATWER
GS	-GLYCOL SUPPLY
GR	-GLYCOL RETURN

LETTER DESIGNATION OF EQUIPMENT TYPE	
A	AGITATOR
B	BLOWERS, FANS, COMPRESSORS
C	COLUMN
D	DESSICANT DRYERS
E	HEAT EXCHANGERS, BOILERS, COOLING TOWERS
F	FILTER
G	
H	HOODS
I	INCUBATORS
J	
K	CHILLERS
L	
M	MIXER
N	
O	REVERSE OSMOSIS
P	PUMP
Q	
R	FERMENTOR
S	DISTILLATION UNITS
T	TANKS AND VESSELS
U	ULTRAFILTRATION
V	
W	SCALE
X	MISCELLANEOUS
Y	
Z	ROLLER BOTTLE RACKS

LETTER IDENTIFICATION OF INSTRUMENTS					
	FIRST LETTER		SUCCEDING LETTERS		
	MEASURED OR INITIATING VARIABLE	MODIFIER	READOUT OR PASSIVE FUNCTION	OUTPUT FUNCTION	MODIFIER
A	ANALYSIS	ALARM	ALARM		
B	BURNER FLAME		USER'S CHOICE	USER'S CHOICE	USER'S CHOICE
C	CONDUCTIVITY (ELECTRICAL)			CONTROL	CLOSED
D	DENSITY (MASS) OR SPECIFY GRAVITY	DIFFERENTIAL			
E	VOLTAGE (EMF)		PRIMARY ELEMENT		
F	FLOW RATE	RATIO (FRACTION)			
G	GAGING (DIMENSIONAL)		GLASS		
H	HAND (MANUALLY INITIATED)				HIGH
I	CURRENT (ELECTRICAL)		INDICATE		
J	POWER	SCAN			
K	TIME OR TIME-SCHEDULE			CONTROL STATION	
L	LEVEL		LIGHT (PILOT)		LOW
M	MOISTURE OR HUMIDITY				MIDDLE OR INTERMEDIATE
N	USER'S CHOISE		USER'S CHOICE	USER'S CHOICE	USER'S CHOICE
O	USER'S CHOISE		ORIFICE (RESTRICTION)		OPEN
P	PRESSURE OR VACUUM		POINT (TEST CONNECTION)		
Q	QUANTITY OR EVENT	INTERGRATE OR TOTALIZE			
R	RADIOACTIVITY		RECORD OR PRINT		
S	SPEED OR FREQUENCY	SAFETY		SWITCH	
T	TEMPERATURE			TRANSMIT	
U	MULTIVARIABLE		MULTIFUNCTION	MULTIFUNCTION	MULTIFUNCTION
V	VISCOSITY			VALVE, DAMPER OR LOUVER	
W	WEIGHT OR FORCE		WELL		
X	UNCLASSIFIED		UNCLASSIFIED	UNCLASSIFIED	UNCLASSIFIED
Y	USER'S CHOICE			READY OR COMPUTE	
Z	POSITION			DRIVE, ACTUATE OR UNCLASSIF. FINAL CONTROL ELEMENT	

Pharmacia
Process Separation Division

Drawn KLO	Date 881215
Ck'd JAC	App'd

Title		
LEAD SHEET PIPING AND INSTRUMENTATION		
Scale	Proj No	Dwg No BPE-1005-1

HAND ELECTRICAL STATION DESIGNATIONS	
PB	PUSHBUTTON
PBL	PUSHBUTTON AND LIGHT
PB^2	TWO PUSHBUTTONS
PB^2L	TWO PUSHBUTTON AND LIGHT
PB^2L^2	TWO PUSHBUTTONS WITH 2 LIGHT
SS	SELECTOR SWITCH
SSL	SELECTOR SWITCH AND LIGHT
SSL^2	SELECTOR SWITCH AND TWO LIGHTS
HOA	SELECTOR SWITCH/THREE POSITIONS (HAND-OFF-AUTO)
OCA	SELECTOR SWITCH/THREE POSITIONS (OPEN-CLOSE-AUTO)
HOR	SELECTOR SWITCH/THREE POSITIONS (HAND-OFF-REMOTE)
OC	SELECTOR SWITCH/TWO POSITIONS (OPEN-CLOSE)
OCO	SELECTOR SWITCH/THREE POSITIONS (OPEN-CLOSE-INTERLOCK OVERRIDE)
KS	KEYED SWITCH

INSTRUMENT POWER SUPPLY ABBREVIATIONS	
AS	AIR SUPPLY
ES	ELECTRIC SUPPLY
GS	GAS SUPPLY
HS	HYDRAULIC SUPPLY
NS	NITROGEN SUPPLY
SS	STEAM SUPPLY
WS	WATER SUPPLY

FUNCTION DESIGNATIONS

* Used for single-input relay.
| Used for relay with two or more inputs.

SYMBOL	FUNCTION	NOTE
1. 1-0 or ON-OFF	Automatically connect, disconnect, or transfer one or more circuits provided that this is not the first such device in a loop.	
2. $\sum$ or ADD	Add or totalize (add and subtract)	|
3. Δ or DIFF	subtract	|
4. ± + [-]	Bias	*
5. AVG	Average	
6. % or 1:3 or 2:1 (typical)	Gain or attenuate (input:output)	*
7. [×]	Multiply	|
8. ÷	Divide	|
9. $\sqrt{}$ or SQ. RT.	Extract square root	
10. x^n or $x^{1/n}$	Raise to power	
11. f(x)	Characterize	
12. 1:1	Boost	
13. [>] or HIGHEST (MEASURED VARIABLE)	High-select. Select highest (higher) measured variable (not signal unless so noted)	
14. [<] or LOWEST (MEASURED VARIABLE)	Low-select. Select lowest (lower) measured variable (not signal unless so noted)	
15. REV	Reverse	
16. a. E/P or P/I (typical)	Convert For input/output sequences of the following: DESIGNATION / SIGNAL E Voltage H Hydraulic I Current (electrical) O Electromagnetic or sonic P Pneumatic R Resistance (electrical)	
b. A/D or D/A	For input/output sequences of the following: A Analog D Digital	
17. $\int$	Integrate (time integral)	
18. D or d/dt	Derivative or rate	
19. 1/D	Inverse derivative	
20. As required	Unclassified	

Figure 28. A piping and instrument (P&I) drawing of a complex chromatography system.

Defining a good functional specification requires understanding both the overall process and the biochemistry involved. For example, the optimal choice of in-line filters depends on the early downstream procedures (e.g. centrifugation and filtration) and the hydrophobic characteristics of the sample. Proper filters for continuous in-line filtration are determined by long-term laboratory scale trials using different filter types. The choice of the proper in-line filter can be an important part of process development, and will have an impact on process performance, time consumption, and filter consumption in production. In-line

	PTFE	ETFE	CTFE	PFE	PVDF	PEEK	PI	PP	EPDM	Titanium	Platinum	Glass (Borosilicate)
Acetic acid (10%)	+	+	+	+	+	+	0	+	+	+	+	+
Acetone	+	+0	+	+	0	+	+	+		+	+	+
Acetonitrile	+	+	+	+	02	(+)		(0)		+	+	
Ammonia (aqueous)	+	+		+		+		+		+	+	+
Ammonium sulfate	+	+	+		+			+		+		
Beer	+	+				+		+		+	+	+
Chloroform	+	+0	+0	+0	+0	+		0–		+	+	+
Citric acid	+	+	(+)		+	+		+		+	+	+
Cyclohexane	+	+		+	+	+		+0	2	+	+	
Ethanol	+	+	+	+	+	+		+	+1	+	+	
Ethylene glycol	+	+	+	+	+	+		+	+			
Formic acid	+	+	+	0–	+	+		+	+2	0	+	+
Glycerol	+	+	+	(+)	+			+	+			
Guanidinum hydrochloride	+	+						+	+			
Hexane	+	+0	+0	+0	+0	+		+0	–2	+		
Hydrochloric acid (10%)	+	+	+	+	+	+	(+)	+	+	0	0	+
Hydrochloric acid (concentrated)	+	+	+	+	+	+		+		–	0	+

Methanol	+	+	+	+	+	+		+	+1	+	+	+
Methylene chloride	+	+0	(+0)	(+0)	0	+		0−	03	+	+	
Nitric acid (<25%)	+	+	+	(+)	+	+0	0	+	+	+	+	+
Nitric acid (<50%)	+	+	+	(+)	+	+0	(0)	0		+	+	+
Phosphoric acid (20%)	+	+	+	(+)	+	+		+	+	+	+	+
Phosphoric acid (50%)	+	+	+	(+)	+	+		+	+	0	+	+
Potassium carbonate	+				+	+		+		+		
Potassium chloride	+				+	+		+		+		
Propranol	+	+		(+)	+			+	+			
Sea water	+	+	+	(+)	+	(+)	+		+	+	+	+
Sodium acetate	+	+			+	+		+		+		
Sodium carbonate	+	+	+		+	+	0	+		+		
Sodium chloride	+	+	+		+	+		+	+	+		
Sodium hydroxide (20%)	+	+	+	(+)	0	(+)	0	+	+	+		
Sodium hydroxide (80%)	+	+	(+)	(+)	0	(−)		+		−		
Sodium sulfate	+	+	+		+	+		+		+		
Sulfuric acid (<50%)	+	+	+	+	+	+	0	+	+	−	+	+
Tetrahydrofurane	+	+0	0	+0	0+			0−	−	+		
Toluene	+	+0	+0	+0	+		+	0−	−2	+		
Trifluoroacetic acid (30%) (TFA)	+	+0	0	0				(+)		0		
Triton X-100	+	+	+	+	+	+		+	+			
Urea	+		(+)		+	+		+	+			

Figure 29. Chemical resistance chart for selected materials used in process chromatography. Primary system—broad classification: −, not resistant; 0, limited resistance; +, resistant; space, no data available; parentheses, resistance is conditional or the data is from a limited test series. Secondary system—resistance qualification: 0, limited resistance; 1, fair resistance; 2, good resistance; 3, very good resistance.

filters affect the life of separation media and as a result, the economics of the entire process.

Establishment of the functional specifications leads to a piping and instrument (P&I) drawing that includes all instruments and process functions (see *Figure 28*). But before it is possible to choose the appropriate equipment for the defined functions, the chemical and physical specifications of the system must be defined.

Chemical specifications

The buffers defined by the purification protocol influence what construction materials are allowed. For example, even common NaCl buffers at mildly acidic conditions can cause corrosion problems with stainless steel. The sample can also limit the choice of suitable construction materials if the protein of interest is sensitive to contact with certain surfaces. The choice of construction materials may be further restricted by the need for regular cleaning of the system with strong solutions. (See also Chapter 9, Process Hygiene.)

The chemical resistance chart shown in *Figure 29* gives guidelines on the chemical resistance of different plastics, elastomers, metals and glass. It is important to note that 'resistance' is not an absolute but a relative term. Plastics and elastomers, in particular, may be affected by long-term exposure to chemicals. This is not found in standard tests, simply because the duration of the tests is limited. Furthermore, when under pressure, plastics and elastomers may be considerably more affected by chemicals. Extensive tests, therefore, should be performed to verify leakage and resistance of the material to the exact combination of chemicals, pressure and temperature used in the process.

For production scale chromatography of biologicals, a general rule of thumb is to use only glass, fluoroplastics and stainless steel. This gives full freedom in the choice of the separation chemistry and sterilization techniques. (However, fluoroplastics may not be mechanically suitable for sealing components, and glass can only rarely allow for in-line steam sterilization.) For further information, see Refs 4 and 5.

Pressure specifications

Pressure specifications are estimated from laboratory measurements of the pressure drop over the packed chromatography column and information from equipment suppliers showing the pressure drops over the liquid handling equipment and the empty column. Pressure measurements of the assembled system must then be made to determine the specifications. The pressure drop information should be shown as a function of flow rate. The operating system pressure is calculated as the sum of the pressure drop over the packed column and the equipment to be installed between the pump and the fraction collection system.

With separation media smaller than 30-40 μm, the contribution from the extra-column equipment to the overall pressure specification is defined largely by the column back pressure.

Hygienic design

Solvent compatibility of equipment and the hygienic design of chromatographic systems have become more important with the introduction of separation media that withstand strong alkaline solutions suitable for cleaning-in-place (CIP). (For further details see Chapter 9.)

In principle, it is possible to sterilize a chromatographic system in-line. In reality, however, the geometry of the equipment may not allow the cleaning agent to flush all parts of the system. Equipment should be chosen that does not create stagnant zones. For example, the geometry of a membrane valve allows for free liquid flow over the whole internal surface. But it is not always possible to find hygienically designed equipment. For example, a small scale UV monitor flow cell with normal types of non-sanitary, threaded connections might be needed in the system. It is still possible to achieve good process hygiene, however, by regularly opening the flow cell and cleaning it outside the system.

Good hygiene in process chromatography depends on using hygienically designed equipment where available, sterile-filtered solutions coming into the system, a method for in-line cleaning, and routines to take care of the parts of the system with geometry that makes in-line cleaning less efficient.

SELECTION OF COMPONENTS

A chromatography system can be quite simple, consisting of only a few components, or very complex. In addition to the automation hardware and software described below, a complete system includes: columns, valves, pumps, monitors and sensors, tubing or piping, and fraction collectors. For further reading, see Refs 6 and 7.

Columns

A chromatography column is designed to create as little zone spreading as possible and to allow for packing the chromatography

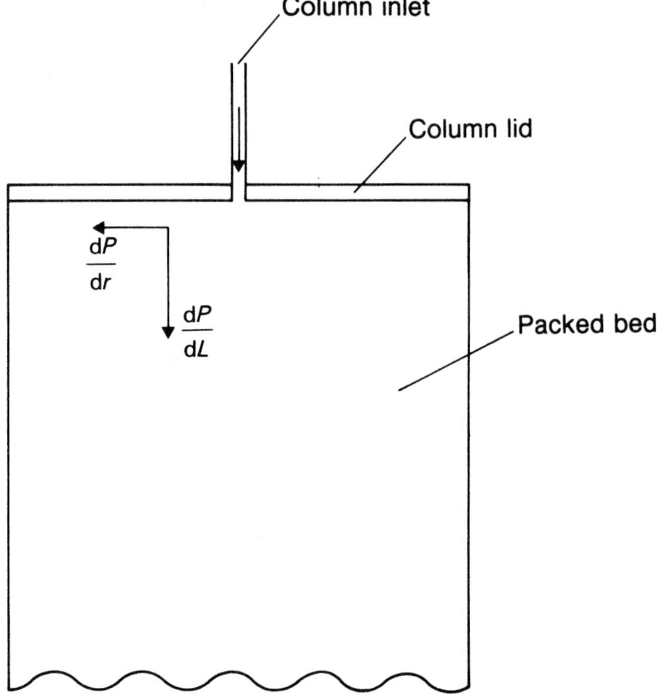

Figure 30. Radial and axial back-pressure in a column distribution system.

media to give the highest efficiency. Minimal zone spreading is achieved by making flow distribution as even as possible over the column inlet and outlet distribution systems. Technically, the construction can vary, but all columns showing an even flow distribution have a radial pressure drop which is negligible in relation to the axial pressure drop in the inlet (see *Figure 30*).

One commercially available distribution design consists of one central inlet with anti-jet devices to prevent the incoming liquid from disturbing the packed chromatographic media. This simple construction, combined with a small chamber just under the net to create a low radial back pressure, has given good results in columns with diameters up to approximately 20 cm.

For columns with larger diameters, a number of constructions are available. A multiple-inlet flow distribution system has been used in the chromatographic purification of insulin (see *Figure 31*).[8] A circular inlet equipped with a circular anti-jet plate has been shown to be as effective as a multiple-inlet design. And the single-inlet connection improves the hygienic design of the column. These two inlet designs are suitable for *low pressure* chromatography with relatively large-particle separation media (larger than 30 μ).

When operating with smaller particles that create higher back pressure, simple 'one-layer nets' that keep the packed

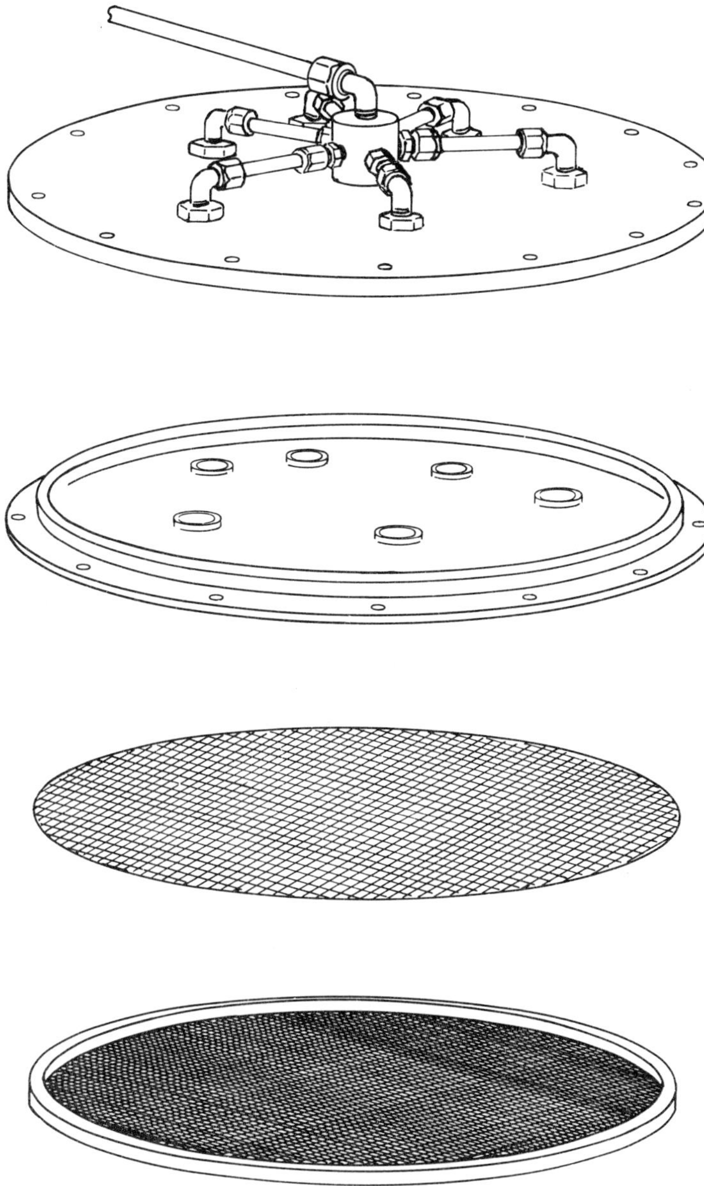

Figure 31. The Pharmacia Column KS370 distribution system.

chromatographic media in place are not commercially available. Fine-meshed depth filters prevent particles from escaping from the column or from simply blocking the nets. Depth filters, however, have a disadvantage in protein processing since the relatively large filter surface can become blocked due to adherence of molecules in the feed to the filter material. In continuous production situations, this drawback of depth filters might create serious problems. For some feeds, it may be necessary to use chromatographic media with larger particles or, alternatively, use a column that allows frequent and simple filter changes.

The choice of column depends on the chemical and physical specifications and, most importantly the separation media. For low pressure chromatography (up to 3 bar), there are now columns commercially available constructed of plastic, glass or stainless steel. These columns range in size from laboratory scale up to diameters of approximately 200 cm.

Traditionally, biochemists have preferred columns that allow for visual inspection of the packed media. But it is useful to keep in mind that as the column diameter increases, a smaller portion of the packed media can be seen. Both glass and some plastics allow for visual inspection, but some plastics do not meet pharmaceutical industry requirements for chemical leakage, hygienic design and in-line cleaning. Columns that allow for high flow rates and meet pharmaceutical industry standards are constructed of glass or high quality stainless steel.

As mentioned above, occasionally a normal stainless steel construction might not be compatible with the chemical environment. In this case, fluoroplastic-coated stainless steel constructions are recommended. Alternatively, columns constructed of higher quality stainless steel that is unaffected even by strongly acidic halide-containing buffers can be employed.

For *medium pressure* chromatography (up to 20–40 bar), columns constructed of glass or stainless steel are available in diameters up to 60–100 mm. Diameters greater than 60–100 mm require stainless steel.

For *high pressure* process chromatography, the only practical construction material is stainless steel. Unfortunately, as the system pressure increases, so does the column cost. At high pressures, the column may become classified as a pressure vessel. Pressure vessel codes differ from one country to another and must be defined before choosing a column for production purposes.[9,10]

Table 15. Valve functions.

Automatic buffer switch
Fraction collection
Filter by-pass
Column by-pass
Pump switch
Forward or reverse flow through the column
Connecting columns in series or in parallel
Selecting on-line monitors
Recycling product
System drainage
Cleaning-in-place (CIP) routines

During process development, the use of bed height adaptors allows for evaluation of the optimal bed configuration and simplifies the frequent repacking needed when comparing different separation media. In production, however, a column with fixed end pieces becomes a more hygienic and economical alternative if frequent repacking is not required.

Valves

Valves are used for directing liquid through the chromatography system, and valve functions are shown in *Table 15*. These functions can be either manual or automatic, or both.

For buffer selection and fraction collection, multiport valves are preferred to minimize dead volumes that provide a location for bacterial growth. Filter and column by-pass are best achieved by manifolding two-way valves to cope with both differential pressure and hygienic aspects. Three-way valves are normally less suitable in these functions because of the differential pressure and hygienic conditions. Ball valves, despite being well suited for by-passing a flow, are not considered hygienic.

For other valve installations, two-way valves are generally the most acceptable because of good hygienic design and general applicability. The two-way valve can be operated either electromagnetically or by compressed air. The only general drawback of the two-way valve is the dead volume created by the installation.

Depending on the scale of operation, system pressure, and chemicals used, diaphragm valves are the valves of choice for hygienic, reliable, production scale chromatography. For further information on valve design, see Ref. 7.

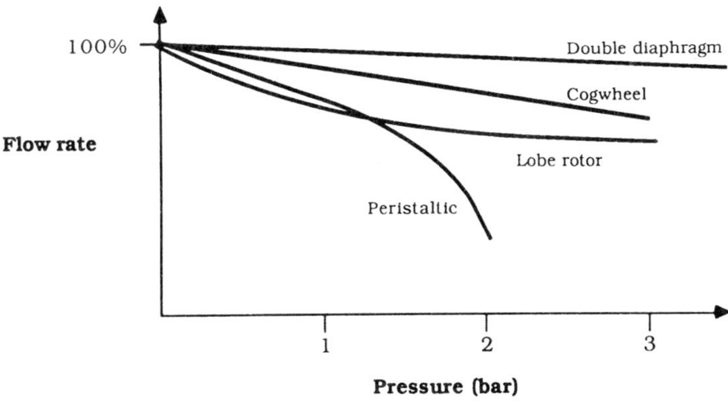

Figure 32. Pressure/flow rate curves for different types of pump.

Pumps

When selecting a pump for process chromatography, hygiene, chemical resistance, pressure/flow rate capabilities, temperature tolerance, shearing, speed control and pulsation should be considered.

Hygiene

For production scale liquid chromatography, the pump should be easily disassembled for cleaning and permit steam sterilization and cleaning-in-place. There should be no stagnant zones in the pump where microbial growth may occur.

Chemical resistance

The materials in the pump must be compatible with cleaning procedures and chemicals used in the process. The number of suitable pumps is rather limited, particularly if steam sterilization is to be used. It is important to know what materials are used in the pump seals. Many chromatographic processes require EPDM or fluororubbers because of their chemical resistance.

Pressure/flow rate

It is most important that the pump covers the entire flow rate range of the process and that the discharge pressure is adequate over the whole range (*Figure 32*).

Temperature tolerance

The operating temperature range of different pumps varies. Usually, this does not cause any problems because most chromatographic processes are run at room temperature. However, if steam sterilization or hot water is used for sanitization, considerable care must be taken in selecting the appropriate pump.

Shearing

Most proteins are sensitive to shearing. When choosing a pump for protein processing, this is one of the most important aspects to consider. Several commonly used industrial pumps, e.g. centrifugal gear and screw pumps, cause protein shearing. Eriksson[11] has reported up to 60% breakdown of albumin due to pump shearing. A detailed description of the selection and use of peristaltic pumps for production chromatography is given by Brown.[12]

Speed control

Liquid chromatography systems are normally operated over a wide flow-rate range. During elution the flow rate is usually considerably lower than during equilibration and cleaning steps. To enable operation over the entire flow-rate range, it is necessary to have pumps with automatic speed control.

Pulsation

Pump pulsation should be minimized to prevent disturbance of the packed column. If a pulsation pump must be used, care should be taken to keep the pulsation volume low. Air traps positioned after the pump act as pulsation dampeners.

Monitors

The most frequently used monitors for liquid chromatography measure UV, conductivity and pH. Flow meters, pressure sensors and air sensors are also used quite often. The recorded measurements become part of the batch documentation.

UV monitors

UV monitors measure the absorbance of UV light. Usually the UV absorption of proteins is measured at a wavelength of 280 nm. Multiple wavelength and scanning wavelength monitors may be useful for laboratory and even pilot scale chromatography. In production chromatography, single-wavelength monitors are preferred because of their reliability and simplicity. Furthermore, little of the available scanning equipment is designed for industrial use.

There are functions that are advantageous in both laboratory and process chromatography UV monitoring, i.e. auto-zero base line and event mark. For process chromatography, there are additional features to be considered. These include: flow-through cell design, variable path lengths, sanitary connections and cell interior, nonsensitivity to electrical noise, lamp surveillance, explosion protection and adjustable alarm levels.

Table 16. Possible variations in pH and ionic strength (*I*) in buffers used in the different chromatographic steps.

	Correct		Range	
Step	pH	*I*	pH	*I*
Buffer exchange	7.0	0.025	Not critical	0.023–0.027
Anion exchange				
Buffer 1	5.20	0.025	5.10–5.25	0.023–0.027
Buffer 2	4.50	0.025	4.40–4.60	0.025–0.030
Buffer 3	4.00	0.15	Not critical	Not critical
Cation exchange				
Buffer 1	4.50	0.025	4.30–4.70	0.023–0.027
Buffer 2	5.50	0.11	5.45–5.55	0.11–0.12
Buffer 3	8.00	0.40	Not critical	Not critical
Gel filtration		0.05 M NaCl	Not critical	Minimum 0.02

Conductivity monitors

Conductivity monitors measure ionic strength. They can be used to monitor and document conditions during elution. They are also useful for monitoring and automating cleaning and equilibration steps. It is important to keep in mind that both conductivity and pH (see below) measurements are temperature dependent.

Industrial conductivity monitor requirements are, in many respects, similar to those described above for process scale UV monitors. Since conductivity monitors are used for water purification systems and waste water treatment units, there are plenty of large scale flow jackets available. Little has been done in the past to make the cells and flow jackets sanitary or to minimize the internal volume of the flow jackets. Fortunately, recent demands from the biotechnology industry have led to the development of new flow-through cells.

pH monitors

In many applications, accurate pH control is critical to the success of the separation. Minor fluctuations in the pH might ruin the separation since some proteins elute within 0.1 pH unit of each other. Furthermore, an equally small variation in pH can affect the solubility of a protein and can lead to column blockage. *Table 16* shows a pH sensitivity investigation on a process for separation of human serum albumin (HSA) from blood plasma. The results indicate that the pH during part of the process may not vary more than 0.05 pH unit to achieve reproducible results.[13] From a technical point of view, this is a very narrow limitation.

Industrial pH monitoring is very difficult, particularly because of the sensitivity of pH electrodes to fouling. The design of a pH measuring unit, therefore, needs particular attention. Most important, the electrode should be easily removed from its flow jacket for cleaning, calibration, or replacement.

Flow meters

The flow rate in a liquid chromatography system is usually controlled using a flow meter signal as input and a controller to adjust the speed of the pump. It is difficult to find suitable flow meters for flow rate ranges used in process chromatography, particularly since the range may vary by a factor of 10 or more in one process. The flow meter should not be sensitive to viscosity changes, and a sanitary design is preferred. Several types of flow meters are available. These include: turbine, electromagnetic and vortex meters.

Pressure sensors

Pressure sensors monitor the pressure drop over vital parts of the system, and act as an alarm or process control function. There are two types of sensor: a manometer with local indication of pressure (mechanical, with or without alarm function) and electronic pressure transducers. Neither of these are ideal for chromatography as they both increase dead volumes in the flow path. Recent developments indicate that this handicap may be soon overcome.

Air sensors

Air sensors are used to protect the chromatography column from air and to enable complete emptying of the sample tank. The column can also be protected by an air trap. Usually an air trap and an air sensor are used in series.

The complete emptying of the sample tank is essential if the sample has a high value. Level sensors can be used to achieve sample tank emptying, but the air sensor has an advantage over level sensors. Since the air sensor is usually placed on the outlet line, it is not in direct contact with the tank contents.

Contact with the fluid stream can be avoided completely. The piece of tubing monitored passes through the sensor without the process fluid coming in contact with the sensor. Because of the operating principle, the tubing passing through the sensor cannot be conducting, which makes stainless steel pipe unsuitable for use with air sensors.

In stainless steel systems, plastic tubing is substituted for stainless steel where the air sensors are installed.

Air sensors usually react to changes in capacitance and can only be used for solutions with a conductivity greater than 0.3-0.5 $mS\,cm^{-1}$. Air sensors are unsuitable for detecting air in distilled water.

Tubing or piping

Buffers and sample are pumped from their respective containers to the chromatography column in tubing or piping. The acceptable pressure drop in a chromatography system is determined by the discharge pressure of the pump in the system and the maximum operating pressure of the respective components in the system, including the tubing. The system should be designed in such a manner that most of the pressure drop is produced by the separation media in the column. As a general guideline for low pressure chromatography, the maximum pressure drop in the tubing should be 10-15% of the available pressure from the pump. This leaves around 50% of the available pressure for the separation media and the rest for all other components in the system.

When selecting tubing or piping for a chromatography system, the following additional aspects need to be considered: dimensions, internal volume, chemical stability, temperature, flexibility and price.

Fraction collectors

In the laboratory, the fraction collector is just as important as the column. At process scale, other considerations affect the choice of equipment for collecting fractions, the most important being the number of fractions collected. Rarely are more than five or six fractions collected in a well-developed production process. During process development, however, more fractions will usually be collected. Multiport valves or manifolds of two-way valves are the most suitable fraction collection method for production.

AUTOMATION

Automation of a process includes control, documentation and evaluation. This section describes advantages of automation, different control systems, and hardware and software specifications for chromatographic automation systems for development to production stages.

Advantages of automation

There are many advantages to automating chromatographic processes. These include: decreased labor costs, reduced process development time, decreased capital costs, increased reproducibility, increased reliability, improved work environment, improved overview and improved documentation.

One of the most obvious reasons for automating a process is to reduce labor costs. Automation has the added advantages of freeing competent personnel for less tedious tasks and reducing process development time. If the process can run 24 h a day, capital investment costs can be reduced. Since the effect of human error is reduced, automation increases reproducibility. Reliability can be increased by adding check points and automatic alarms; the operator knows if the process is working properly. The check points and automatic alarms can be displayed to allow quick access to all parts of the process. Processes that need to be in cold or hygienic rooms can be remotely controlled, which improves the working environment for the operator and reduces the risk of contamination. For a complex process, automation can improve the overview of the process with logical schematic displays, including set and actual levels of control parameters. Methods, process parameters, and other process information can be stored automatically for each process run.

Control systems

An automation system can be based on different types of control hardware, for example, dedicated controller, programmable logical controller (PLC), personal computer based system, or combinations of these three.

Dedicated control system

This uses hardware optimized for a specific application. This reduces the capital cost because you only pay for what you need. The controller is easy to use and ready for installation. The built-in software is easy to use, i.e. no programming skills are necessary. The dedicated controller can work as a stand-alone or as a remote unit in a computer system. One disadvantage is that expansion possibilities and modifications can be limited. A dedicated controller has its place both in process development and in production plants.

Programmable logical control (PLC) system

A general system that can be used in any type of application. The hardware is often built in a modular style so that the user can change the control capacity by simply adding or replacing modules in the system. The PLC system needs some preparation before the system is ready to use, e.g. assembly of the different modules. Furthermore, extensive programming is necessary in most cases. The PLC system can work as a stand-alone or as a remote unit in a computer system. Those PLC systems that are designed for process environments are suitable for production.

Personal computer system

Another general system that can be used in any type of application. Often the computer communicates with a remote unit, i.e. a dedicated or programmable logical controller. Alternatively, I/O interfaces are installed in the computer or in an interface box with no intelligence. In control systems consisting of computers combined with remote

Table 17. Automation for process development and production systems: hardware specifications.

Compatibility with industrial environment
- voltage fluctuations
- static electricity
- high electrical noise levels
- humidity
- dust, dirt
- explosion proof, where appropriate

Display of relevant information
- valve status
- process step
- flow rate
- pressure
- monitor signals (chromatograms)
- flow sheet

Number of I/O (digital and analog)

Documentation
- printer/plotter

Automatic alarm handling
- differential pressure over column and filters
- high and low pressure levels
- high and low pH levels
- high and low conductivity levels
- high and low flow rates
- high and low tank levels
- air detection
- monitor malfunction (especially UV monitor lamp)
- valve malfunction
- monitor malfunction
- documentation of all alarms

units, the latter control the process, and the computer documents, evaluates and manages one or more systems. In the case of I/O interfaces, the computer itself controls the process. In this case, the computer must have software with multi-tasking and real time capabilities. One disadvantage is that this system is not ready to be used. Extensive programming is often necessary to get the system running. There are many software programs available that communicate with laboratory equipment, i.e. computer packages with intelligent front ends. However, for production purposes there are far fewer ready-to-use systems available. Home-made programs for a personal computer are not to be recommended for production, since this type of program usually tends to have a certain 'personality' and production can become dependent on one single person, i.e. the originator of the program. In the relatively immature biotechnology industry, this has actually happened. A complete purification plant had to be rebuilt because the originator of the software (who was not even a programmer) left the company.[14] For further information on software package selection, see Ref. 15.

Hardware and software specifications

Different stages in a project need different types of automation. In the development phase, for example, a more flexible automation system is needed. In production, reliability is more important.

The hardware specifications for an automation system are basically the same for process development as for production. Any of the control systems above could meet the specifications shown in *Table 17.*

The hardware capacity required varies with the complexity of the process. In process development there is a need for expansion possibilities, while in production the system is often fixed. There are

Table 18. Automation for process development and production systems: signal specifications.

Outputs
- 1–3 system pumps, analog control
- 5–40 ON/OFF*

Inputs
- 1–2 flow meters, analog input
- 1–3 monitors, analog input
- 10–40 digital inputs for alarm purposes in production systems
- 1–2 digital inputs per digital output to be recorded

Control loops
- 1 proportional integral (PI) control loop per pump

*For automatic CIP routines and extensive fraction collection, this number of digital outputs increases dramatically.

many signals that need to be included in the hardware. These are shown in *Table 18.*

The software specifications for an automation system for process chromatography depend on which stage the process is in, i.e. process development or production.

In process development, the same chromatography system can be used for many different processes, and the methods can vary to a great extent. Many functions need to be included in the software (*Table 19*).

In production, the automation is used for a well defined process. The software requirements are shown in *Table 20.*

The optimal automation system for process chromatography is a system that can be used in process development, pilot runs and full production. The time required to test and evaluate control of the process will then be minimized.

VALIDATION OF PRODUCTION CHROMATOGRAPHY

In a manually operated system, malfunctions occur primarily as a result of operator error. As the level of complexity increases, the chances of error also increase. The level of validation must then also increase to provide evidence that a process is consistently producing a product meeting the designed quality characteristics. Validation, as defined by the US FDA,[16] means "establishing documented evidence which provides a high degree of assurance that a specific process will consistently produce a product meeting its predetermined specifications and quality attributes." The equipment must, therefore, be designed so that product specifications are consistently achieved.

To properly validate a fully automated system, the function of all components must be tested and documented. Calibration, maintenance

Table 19. Automation for process development systems: software functions.

Flexible number of inputs and outputs

Fast programming by 'non-expert'

Manual control of all outputs during a run

Automatic documentation and display of:
- flow rate
- pressure
- monitor signals
- valve status

Evaluation of results
- resolution
- economy

Table 20. Automation for production systems: software requirements.

Documentation of cost
Documentation of yields
Easy to use for operators
Trend analysis
Manual control
Security, user-defined access to different levels of control
Automatic documentation and display of: flow rate pressure monitor signals valve status
Evaluation of results resolution economy

and adjustment requirements must be determined (see Ref. 16). Safety devices and alarms must be designed so that it is clear that they are functioning properly. The reliability and functional life of monitors should also be known, and precautions should be taken in case of monitor malfunction. The control system itself needs to be well documented. FDA compliance policy guidelines on computerized drug processing are available.[17] For additional information see Refs 18-22.

In summary, selection of chromatography equipment requires careful consideration of the needs for flexibility, documentation, reliability, hygiene, automation and validation. After functional, chemical and pressure specifications are established, components or systems can be selected that meet laboratory, pilot plant and production requirements.

REFERENCES

1. Johansson, H., Ostling, M., Sofer, G., Wahlstrom, H. *et al.* Chromatographic equipment for large scale protein and peptide purification. In *Advances in Biotechnological Processes* (Mizrahi, A., ed.). Alan R. Liss, New York, 1988, Vol. 8.
2. Tomson, J. A vendor's perspective on equipment maintenance. *Pharm. Eng.* **8** (1988) 35-39.
3. Paris, N. A. Instrumentation for high performance liquid chromatography. In *Journal of Chromatography Library* (Huber, J. F. K., ed.). Elsevier, Amsterdam, 1978, Vol. 13, pp. 43-74.

4. Wang, Y. J. and Chien, Y. W. *Sterile Pharmaceutical Packaging: Compatibility and Stability, Technical Report No. 5.* Parenteral Drug Association, Philadelphia, 1984.
5. Cowan, C. T. and Thomas, C. R. Materials of construction in the biological process industry. *Process Biochem.* **23**(1), (1988) 5-11.
6. McCabe, W. L. and Smith, J. C. *Unit Operations of Chemical Engineering.* McGraw-Hill, New York, 1976.
7. Coulson, J. M. and Richardson, J. F. *Chemical Engineering.* Pergamon Press, New York, 1983, Vol. 6, pp. 150-154.
8. Janson, J.-C. and Hedman, P. Large-scale chromatography of proteins. In *Advances in Biochemical Engineering* (Fiechter, A., ed.). Springer-Verlag, New York, 1982, pp. 43-99.
9. *Boiler and Pressure Vessel Codes, Section II: Material Specifications; Section VIII: Stainless Steel; Section X: Plastics.* American Society of Mechanical Engineers, Fairfield, NJ, 1986.
10. *AD Meikblatt N6: Glass; N1: Plastic; W2, W4, W7, and B8: Stainless Steel.* TUV (Verenigung der Technischem Uberwachungs Vereine ev. Postfach 103384, 4300 Essen 1).
11. Eriksson, S. Work from Pharmacia LKB Biotechnology.
12. Brown, P. Peristaltic pumps in low-pressure LC systems. *BioPharmacy*, March 1988, pp. 46-50.
13. Berglof, J. H., Eriksson, S. and Curling, J. M. Chromatographic preparation and *in vitro* properties of albumin from human plasma. *J. App. Biochem.* **5** (1983) 282-292.
14. Howell, J. A., Jones, M. and John, M. Control of a large protein ion-exchange plant. *Separations for Biotechnology* (Verrall, M. S. and Hudson, M. J., eds). Ellis Horwood, Chichester, 1987.
15. Tomlinson, E. A. Automating a manufacturing operation selecting the right software. *Pharmaceutical Technology*, February 1988, pp. 54-58.
16. *Guideline on General Principles of Process Validation.* Center for Drugs and Biologics and Center for Devices and Radiological Health Food and Drug Administration, Rockville, May 1987.
17. *Computerized Drug Processing: Source Code for Process Control Application Programs. Guide 7132b.* April 16, 1987. Office of Enforcement, Division of Compliance Policy, Associate Commissioner for Regulatory Affairs.
18. Masters, G. and Figarole, P. Validation principles for computer systems. FDA, perspectives. *Pharmaceutical Technology*, November 1986, pp. 44-46.
19. PMA Computer Systems Validation Committee. Validation concepts for computer systems used in the manufacture of drug products. *Pharmaceutical Technology*, May 1986, pp. 24-34.
20. *Software Development Activities. Reference Materials and Training Aids for Investigators. Technical Report.* US Dept Health and Human Services, Public Health Service. Food and Drug Administration, Office of Regulatory Affairs, July 1987.
21. Clark, A. S. Computer systems validation: an investigator's view. *Pharmaceutical Technology*, January 1988, pp. 60-65.
22. Hess, A. An integrated approach to validation. *BioPharmacy*, March 1988 pp. 42-44.

9 *Process Hygiene*

In this section, guidelines for process hygiene, techniques for pyrogen removal and microorganism destruction, and recommended cleaning and storage protocols are presented. Good process hygiene not only reduces the risk of contaminating the product with endotoxins or other substances of bacterial origin, but also reduces the cost of chromatography by increasing the life of separation media.

GUIDELINES

Good process hygiene begins with the use of pyrogen-free water and buffers and a filtered air supply. Personnel should wear clean protective clothing and head caps.[1] Pyrogen-free water can be obtained by using reverse osmosis, ultrafiltration or distillation. The water should be of *United States Pharmacopoeia* (USP) quality, i.e. 'water for injection.' All buffers and cleaning solutions should be filtered through 0.45 μ or, preferably, 0.22 μ filters, and, if possible, the sample should also be filtered. All of the components of the chromatographic system that are in contact with the feed stream should be cleaned in place, preferably with sodium hydroxide. Columns and columns systems should be closed and, during downtimes, it is advisable to keep the column storage solution in motion. Slow flow rates and recycling with a filtered solution during downtimes are recommended to reduce the amount of time required for preparing solutions and column storage solution consumption. If a column is to be stored for a long period, ideally, it should be flushed once a week with a cleaning solution.[2] If it is necessary to work in a cold room, humidity should be controlled to inhibit growth of molds.

The hygienic practices used in the chromatographic production of human albumin from plasma for intravenous injection are discussed in detail in Ref. 1. Whereas plasma should be relatively pyrogen-free, recombinant DNA products produced in *Escherichia coli* will have pyrogens as a known contaminant, and thus the removal of pyrogens from the sample becomes a critical goal.

PYROGEN REMOVAL

Fortunately, most pyrogens bind to anion exchangers. Both DEAE-Sephadex A-25 and A-50[3] and DEAE-Sepharose® CL-6B[4] have been shown to remove pyrogens from contaminated solutions. Strong anion exchangers with quaternary amine functional groups have also been used to remove pyrogens during a chromatographic process. Since ion exchange chromatography is usually one of the first steps in a chromatographic purification scheme, pyrogens may be removed early in the process.

If the pyrogens elute from an anion exchanger with the product, then alternative removal methods may be required. Affinity chromatography has been used to remove the product from pyrogens. An affinity medium in which the ligand is a substrate analog has been used to purify pyrogen-free human α-galactosidase.[5] An immunoadsorbent that specifically removes the product from the feed solution is a very rapid and efficient way of removing pyrogens. Others have reported the use of dye-ligand affinity chromatography and lectin affinity chromatography to remove the product from pyrogens. A resin designed for the specific purpose of removing pyrogenic endotoxins is now available. This material consists of polymixin B—a cyclic, cationic, lipophilic peptide antibiotic with a strong affinity for the toxic lipid A portion of endotoxins—bound to a support material such as agarose.[6] This is a fairly mild technique with a very high capacity for the removal of the endotoxins, and it should be a suitable final step in the purification, if it is necessary. (It should be noted, however, that this technique has been reported not to work in many cases. This is believed to be due to the O-polysaccharide type. But polymixin B-agarose has been used to reduce the endotoxin content of a pertussis vaccine 1000-fold.[7])

Leucocytic pyrogen, otherwise known as interleukin-1,[8] is not an endotoxin, but it produces a fever by raising the set-point of the thermoregulatory center. Other non-endotoxin substances may cause the release of leucocytic pyrogen. The product must be tested for the presence of these pyrogenic materials as well as for endotoxins.

The *Limulus* amoebocyte lysate (LAL) test is used to detect subnanogram amounts of endotoxins. Direct pyrogenicity tests are also performed by injecting rabbits with the final product; this test detects nanogram quantities of endotoxins. In addition, to determine whether substances other than endotoxin are causing pyrogenic responses, human blood mononuclear cells can be cultured with the final product and the cell culture fluid injected into rabbits. A fever

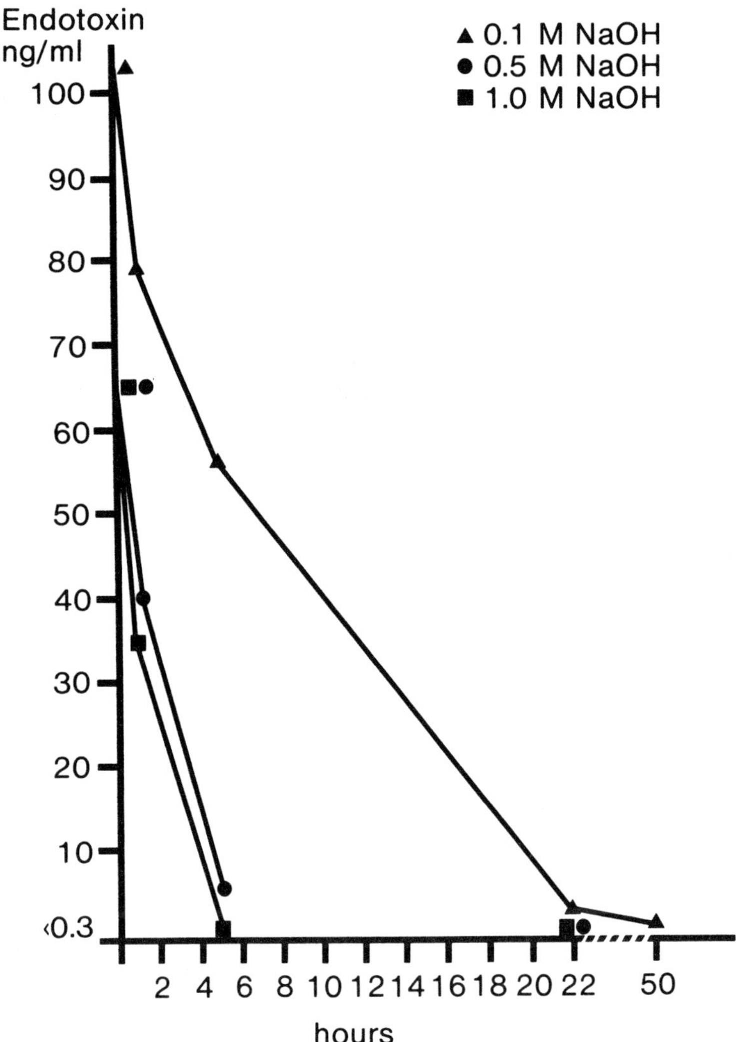

Figure 33. The breakdown of *E. coli* endotoxins by sodium hydroxide at room temperature determined by the chromogenic LAL test (source: work by Pharmacia Biotechnology AB).

in the rabbits indicates that the product contains pyrogenic substances.[9]

Chromatographic media are often, but not always, pyrogen-free. The media themselves do not support microbial growth, but buffers, especially phosphate buffers, do. Therefore, for storage of separation media, alternative solutions such as sodium hydroxide should be used. *Figure 33* shows the reduction of endotoxin levels by sodium hydroxide at several different concentrations and exposure times. An alternative procedure for depyrogenating separation media is given in *Table 21*.[10] In this protocol, 70% alcohol is used to kill the bacteria and 4 M urea solubilizes the pyrogens. This protocol was

Table 21. A procedure for pyrogen decontamination of Sephacryl® S-300. (From Sarafin, T. A., Tsay, K. K., Fluharty, A. L. and Kihara, H. *Biomed. Med.* **28** (1982) 237–240.)

1. 2 V_t 70% EtOH
2. 2 V_t 0.05 M Tris-HCl pH 7.5
3. 1 V_t 4 M urea
4. 3 V_t 0.1 M NaCl

V_t, Column volume.

used for the production of pyrogen-free human arylsulfatase A for therapeutic enzyme replacement trials.

If a gel is already contaminated when purchased or if it becomes contaminated during a cycle, it is necessary to find a method which is cost-effective and safe to use without unpacking the column.

Table 22 shows that sodium hydroxide meets all of these requirements. Several methods for pyrogen inactivation when treating non-protein solutions are discussed in a recent review.[11]

MICROORGANISM DESTRUCTION BY SODIUM HYDROXIDE

The destruction of microorganisms by sodium hydroxide is both time and temperature dependent. *Table 23*[12] shows the time required for destruction of *Bacillus subtilis* spores by sodium hydroxide. Two separation media, S Sepharose® High Performance (30 μ) and

Table 22. A comparison of the advantages and disadvantages of autoclaving, bacteriostatics and sodium hydroxide for chromatography media hygiene.

Autoclaving
- \+ Gives sterility
- − No cleaning effect
- − Does not destroy endotoxins
- − Not possible with packed columns

Bacteriostatics
- \+ Gives a low bacterial level or sterility
- \+ Only possibility with some affinity media
- − Does not remove endotoxins
- − No cleaning effect
- − Risk of traces in the product

NaOH
- \+ Sanitizes complete column systems
- \+ Good cleaning effect
- \+ Destroys pyrogens
- \+ Cannot contaminate the product
- − Careful handling required

Table 23. Times required for 99% destruction of *B. subtilis* spores in sodium hydroxide at different temperatures. (From Whitehouse, R. L. and Clegg, J. *Dairy Res.* **30** (1963) 315–322.)

Molarity of NaOH	20°C	4°C
1.0 (4%)	2 h	22 h
0.5 (2%)	8 h	76 h

Q Sepharose® Fast Flow (90 μ) were challenged with 10^5–10^8 microorganisms per milliliter. The bacterial strains tested included *Staphylococcus aureus, E. coli* and *Pseudomonas aeruginosa.* Also tested were the yeast *Candida albicans* and the filamentous fungus *Aspergillus niger.* The results shown in *Figure 34* indicate that the Gram-negative bacteria, *E. coli* and *P. aeruginosa*, are effectively destroyed by even 0.01 M sodium hydroxide. Gram-positive *S. aureus* showed a much slower reduction rate at this concentration, but a five-fold log reduction was obtained in 60 min with 0.1 M sodium hydroxide. The yeast *C. albicans* showed a rapid decrease in number during the first 1-2 h in 0.1 M sodium hydroxide, whereas a concentration of 0.01 M had practically no effect during the first 24 h. For *A. niger*, a concentration of 0.1 M sodium hydroxide took up to 16 days to cause a five-fold log reduction. But with 0.5 or 1.0 M sodium hydroxide a five-fold log reduction was achieved in 60 min even for *A. niger.* The conclusions from this study are that 0.1 M sodium hydroxide is an effective disinfectant for vegetative bacteria and yeast. Fungal contamination requires a concentration of 0.5–1.0 M sodium hydroxide.

CLEANING PACKED COLUMNS

Media that are compatible with sodium hydroxide are most suitable for production. Some agarose-based ion exchangers have been shown to maintain their separation properties after storage in 2 M sodium hydroxide for 7 days at 40°C.

The cleaning of a packed chromatographic column is normally achieved by first washing the gel bed with a concentrated aqueous solution of a neutral salt such as sodium chloride to remove any weakly bound material. This removes most of the residual proteins and leaves only irreversibly precipitated protein and lipid material. The remaining material can then be solubilized by washing with sodium hydroxide. In the case of hydrophobic interaction or reversed phase media, solvent- or detergent-based cleaning methods should be employed in conjunction with sodium hydroxide. A recommended sanitization protocol for Fast Flow media is shown in *Table 24.*

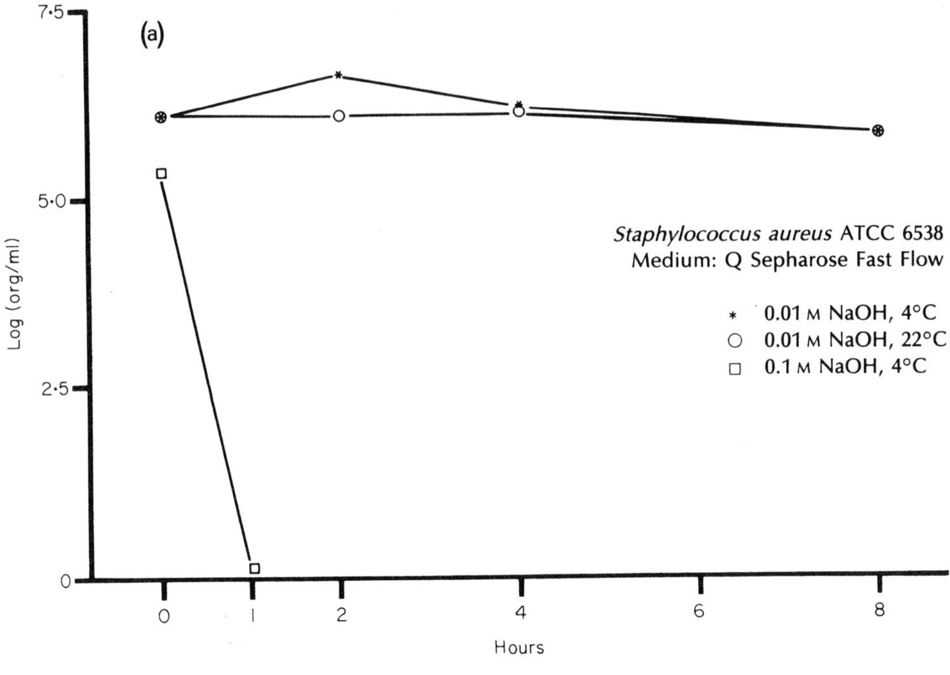

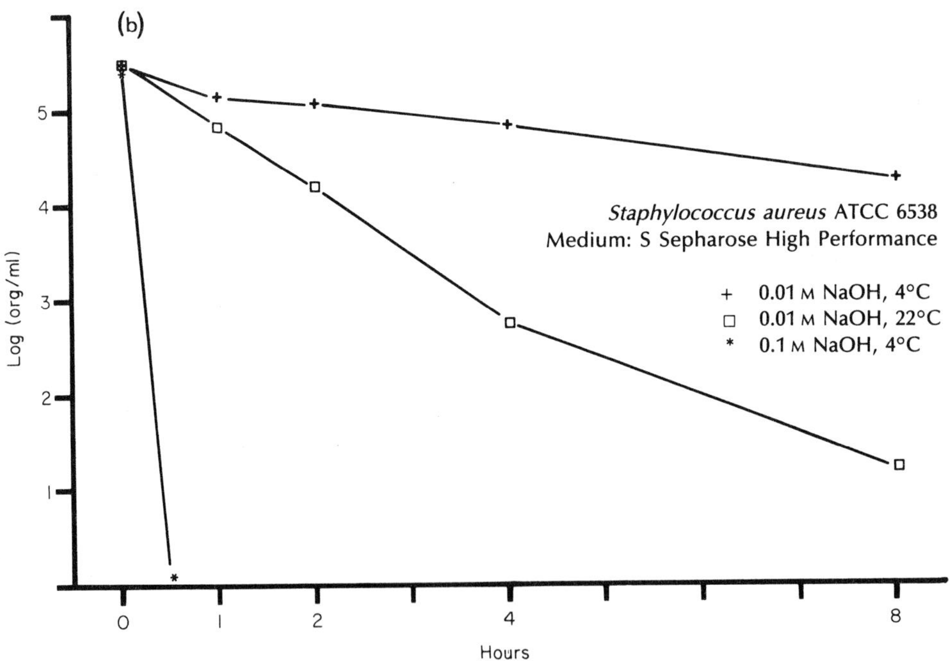

Figure 34 (See page 102 for caption)

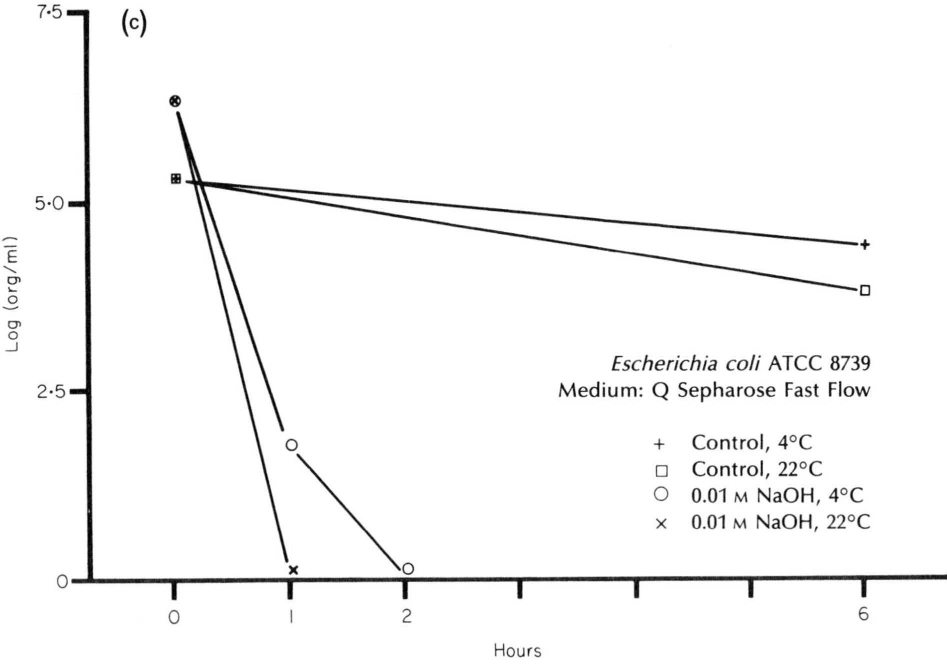

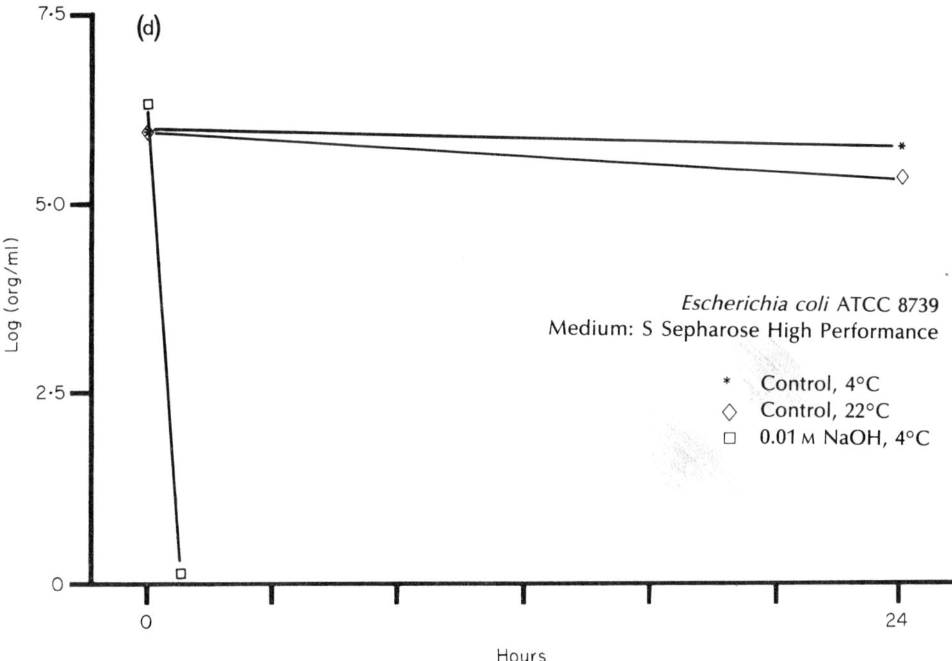

Figure 34 (See page 102 for caption)

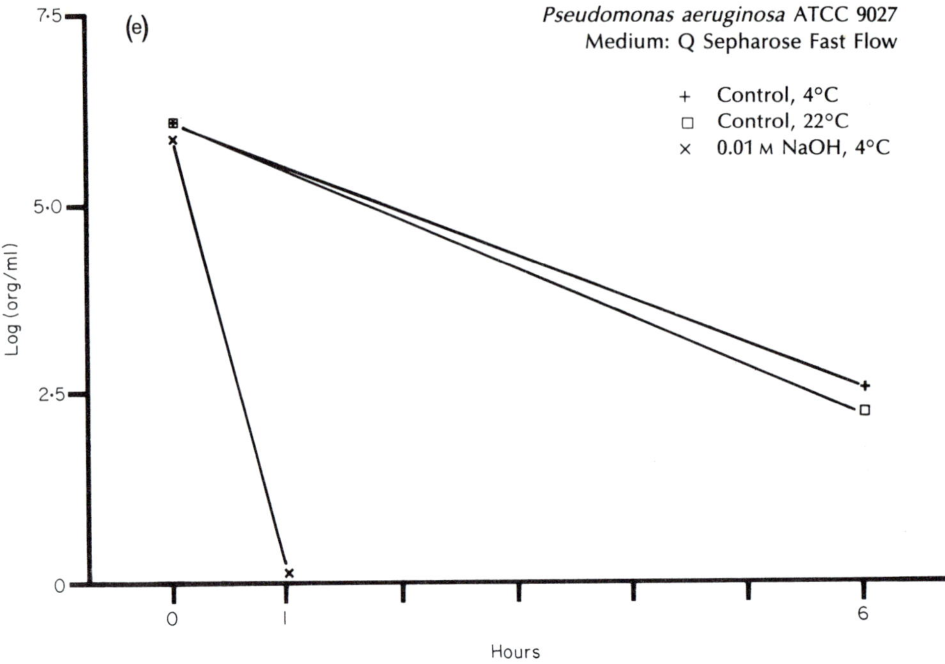

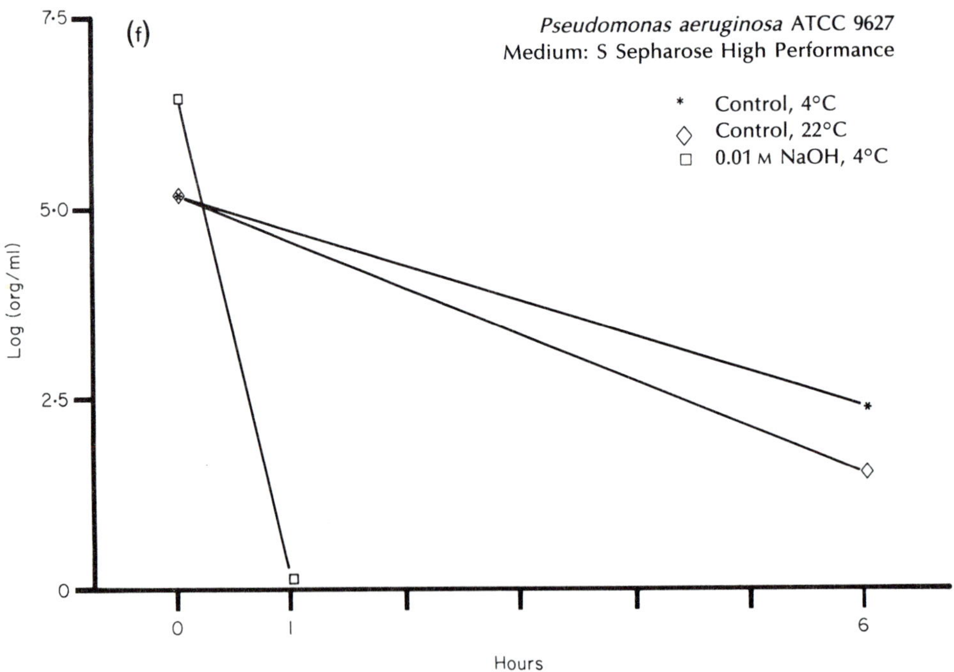

Figure 34 (See page 102 for caption)

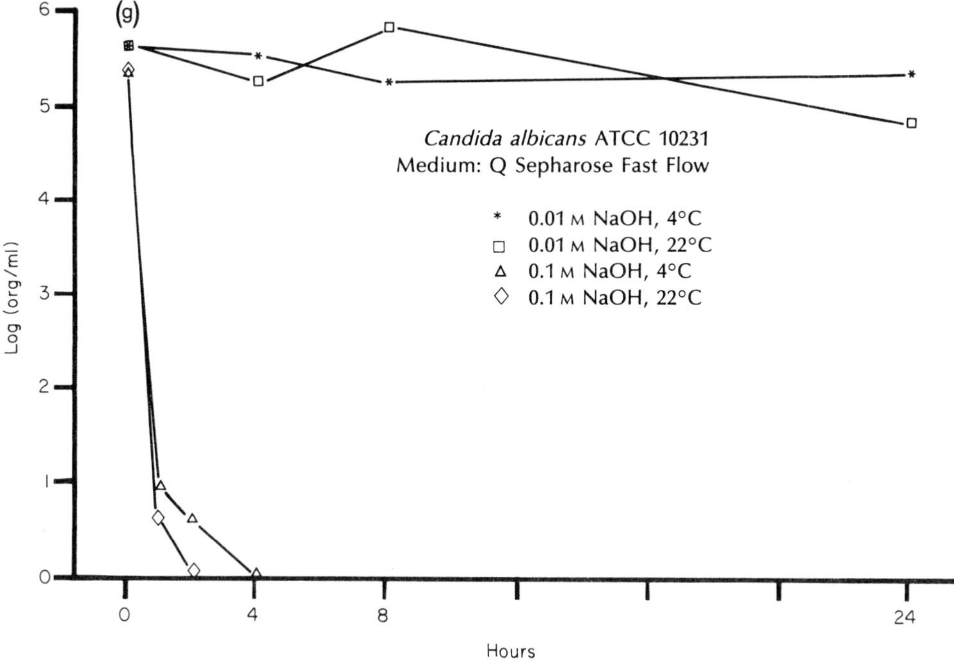

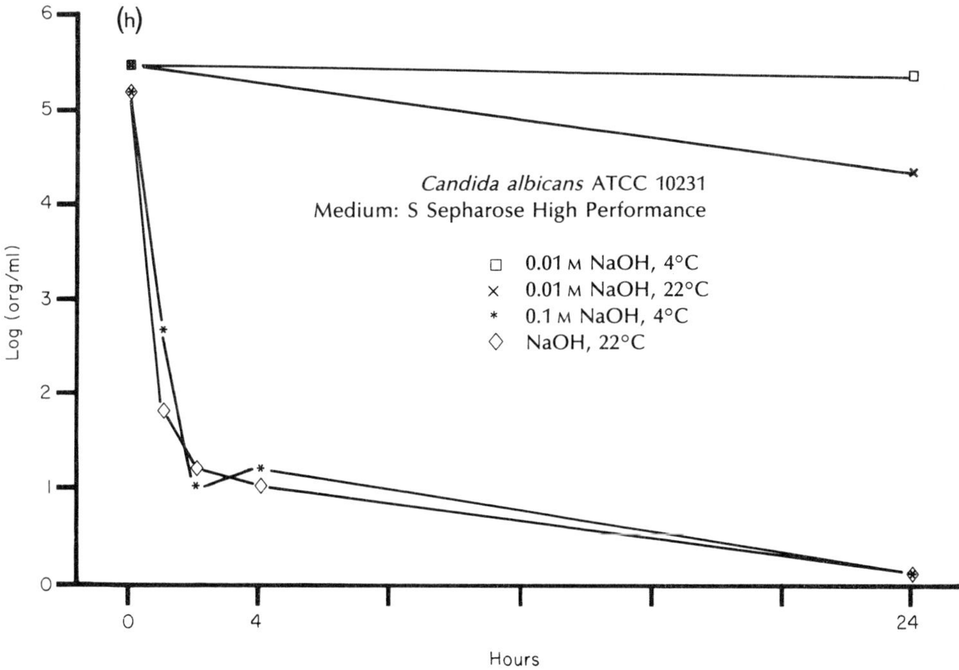

Figure 34 (See page 102 for caption)

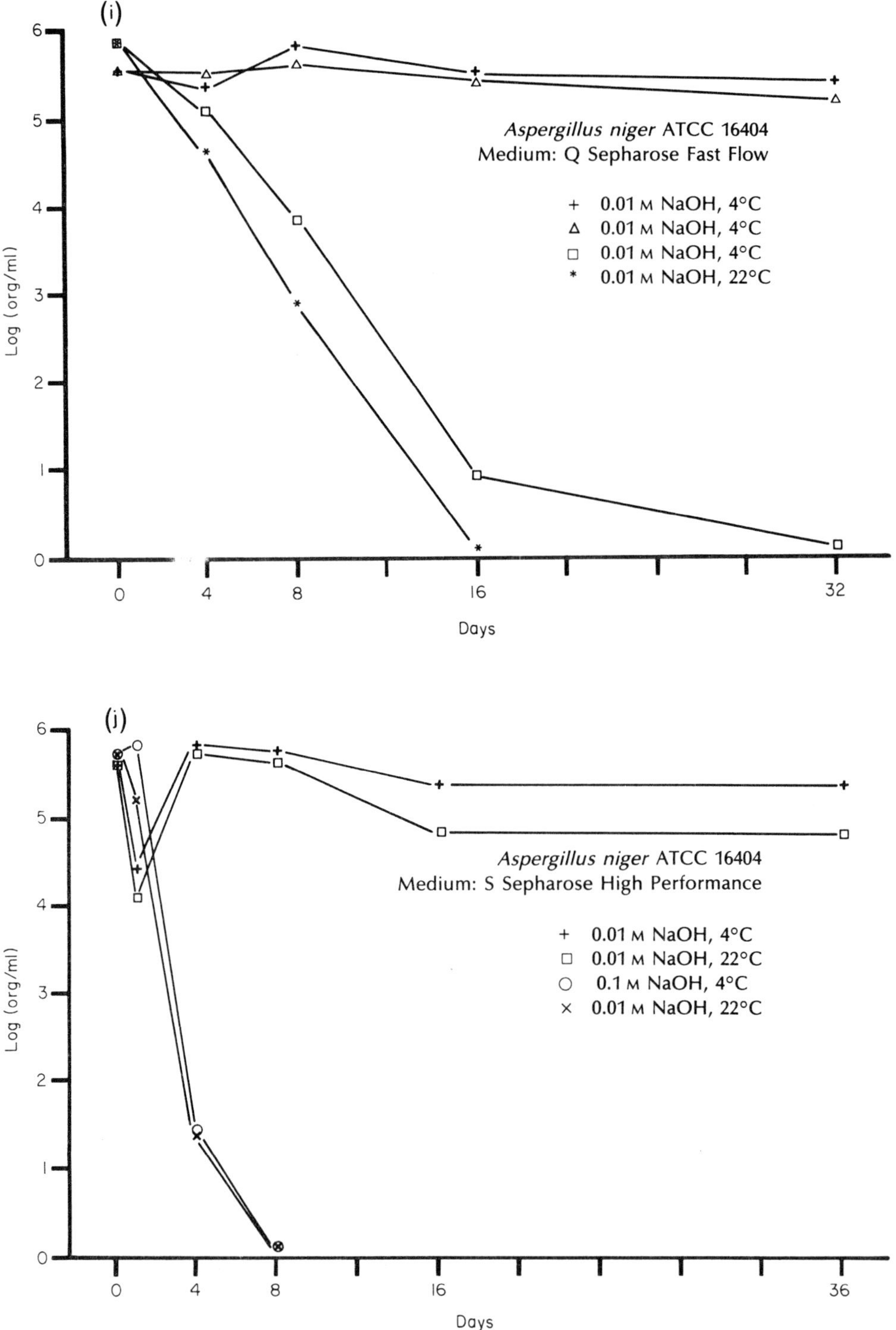

Figure 34. Microbial challenge test on ion exchange media with sodium hydroxide as cleaning agent.

Table 24. Recommended sanitization protocol for Fast Flow media.

Step	Column volumes	Solution	Contact time
(A) Gram-negative bacteria, Gram-positive bacterial (non-sporulating) and yeast			
1	2	Sterile-filtered 2 M NaCl	30 min
2	2	Filtered 0.5 M NaOH	30 min
3	2	Sterile-filtered H_2O*	
4	2	Filtered 0.01 M NaOH	Storage
(B) Molds and spores			
1	2	Sterile-filtered 2 M NaCl	30 min
2	2	Filtered 1 M NaOH	24 h
3	2	Sterile-filtered H_2O	
4	2	Filtered 1 M NaOH	5 h
5	2	Sterile-filtered H_2O*	
6	2	Filtered 0.01 M NaOH	Storage

*Column ready to use after equilibration.

If fragile ligands are used for affinity chromatography in the production of a biological material, it will be more difficult to clean media and maintain them in an aseptic manner. One protocol for cleaning affinity media is given in *Table 25*. Dilute formaldehyde or phenol have been used to sterilize immunoadsorbents, and it was found that the binding capacity remained unchanged by storage in these solutions for periods of 4 weeks at 37°C.[13] However, some ligands may not hold up to these conditions.

It is important to consider process hygiene when optimizing a large scale chromatographic process. The ideal chromatographic medium should be capable of being cleaned in place with sodium hydroxide at high flow rates and, if possible, ligands that support microbial growth should be avoided.

Table 25. Sterilization of affinity media. (From Block, S. S. (ed.). *Disinfection, Sterilization and Preservation*. Lea & Febiger, 1977.)

1. Equilibrate with a buffer consisting of 2% Hibitane digluconate and 0.2% benzyl alcohol
2. Allow to stand for 4 days
3. Wash with sterile buffer
4. Equilibrate with 2% Hibitane digluconate and 0.2% benzyl alcohol
5. Wash extensively with sterile buffer

A cost-effective chromatographic protocol requires good process hygiene to obtain the maximum number of cycles possible. Some chromatographic media used in plasma fractionation plants have lasted up to 5 years. In a production situation, however, the number of cycles one can obtain from a chromatographic medium is more relevant than the lifetime expressed in months or years. Immunoadsorbents have been used for as many as 200 cycles.[14] A Sepharose® ion exchanger has been used for over 1000 cycles in the purification of albumin.[15]

For further reading see Refs 16-18.

REFERENCES

1. Curling, J. M. and Cooney, J. M. Operation of large scale gel filtration and ion-exchange chromatography. *J. Parenteral Sci. Tech.* **36** (1982) 59.
2. Personal communication.
3. Grabner, R. W. US Patent Number 3 897 309, 1975.
4. Shibatani, T., Kakimoto, T. and Chibata, I. Purification of low molecular weight urokinase from human urine and comparative study of two active forms of urokinase. *Thromb. Haemostasis* **49** (1983) 91.
5. Bishop, D. F. and Desnick, R. J. Affinity purification of alpha galactosidase A from human spleen, placenta, and plasma with elimination of pyrogen contamination properties of the purified splenic enzyme compared to other forms. *J. Biol. Chem.* **256** (1981) 1307.
6. Duff, G. W., Waisman, D. M. and Atkins, E. Removal of endotoxin by a polymixin B affinity column. *Clin. Res.* **30** (1982) 565A.
7. Bannatyne, R. M., Jackowski, J. and Cheung, R. Cleaning up pertussis vaccine. *Vaccine* **4** (1986) 91-92.
8. Dinarello, C. A., Bishai, I., Rosenwasser, L. J. and Coceani, F. The influence of lip oxygenase inhibitors on the *in vitro* production of human leukocyte pyrogen and lymphocyte activating factor interleukin 1. *Int. J. Immunopharmacol.* **6** (1984) 43.
9. *Points to Consider in the Production and Testing of New Drugs and Biologicals Produced by Recombinant DNA Technology.* USFDA Recombinant DNA Committee, November 18, 1983.
10. Sarafin, T. A., Tsay, K. K., Fluharty, A. L. and Kihara, H. A Procedure for pyrogen decontamination of Sephacryl S-300. *Biomedic. Med.* **28** (1982) 237-240.
11. Weary, M. and Pearson, F. A manufacturer's guide to depyrogenation. *BioPharmacy*, April, 1988, pp. 22-29.
12. Whitehouse, R. L. and Clegg, L. F. L. Destruction of *Bacillus subtilis* spores with solutions of sodium hydroxide. *J. Dairy Res.* **30** (1963) 315-322.
13. Van Wezel, A. L. and Van der Marel, P. The application of immuno-adsorption on immobilized antibodies for large-scale concentration and purification of vaccines. *Anal. Chem. Symp. Ser.* **9** (1982) 283-292.
14. van der Loo, W. and Hamfers, R. *Protides of the Biological Fluids* (Peeters, H., ed.). Pergamon Press, Oxford, 1976, Vol. 23, p. 603.
15. Viljoen, M., Shapiro, M., Crookes, R., Chalmers, A. *et al.* Large scale recovery of human albumin by a chromatographic method. In *Separation of Plasma Proteins* (Curling, J. M., ed.). Pharmacia Fine Chemicals AB, Uppsala. 1983, pp. 67-72.

16. McCullough, K. Z. Environmental factors influencing aseptic areas. *Pharm. Eng.* **7** (1987) 17-20.
17. Mistalski, T. S. Microbiological contamination troubleshooting. *Pharm. Eng.* **7** (1987) 13-16.
18. Process hygiene in industrial chromatographic processes. *Downstream News and Views for Process Biotechnologists, Pharmacia* No. **2** (1986) 1-3.

10 Economics

The decision to go into production will depend on the overall economy of the process. In this section, we will discuss cost analysis related parameters and total production costs and present an example of cost analysis for some blood-derived products.

Prior to calculating initial investment and production costs, it will be necessary to establish a functional flow sheet. This requires information on the many parameters shown in *Table 26*. These data will be used in dimensioning pumps, valves, columns and tanks, and for calculating the approximate space needed for the whole chromatography installation.[1]

The total cost of producing a biologically active product can be estimated by using equation (1):

$$C_{tot} = \frac{(V_{in} - V_{out}) + \text{process costs}}{\text{product}} \quad (1)$$

where C_{tot}=total production cost, V_{in}=value of product in raw material, V_{out}=value of product in product stream, process costs= chemicals, labor, facilities, and product=yield of product in kilograms. This formula shows that it is essential to have a high yield of product to incur a low cost. Since there is a decrease in yield as the number of chromatographic steps is increased (see *Figure 1*), one way of achieving a cost-effective process is to minimize the number of steps. This can be done by choosing media with a high degree of selectivity, by arranging the chromatography steps in the proper sequence, and by optimizing each step. (*Note*: One genetic engineering company has stated that economic production requires purification yields greater than 25% for eucaryotic proteins produced by mammalian cell culture.[2])

Protein engineering to fuse a 'tag' to the molecule of interest so that it is easily purified may reduce the number of steps required for the purification of recombinant proteins. Polyarginine was fused

Table 26. Parameters to be determined prior to cost analysis.

Gel type
Productivity (g h^{-1} (l gel)$^{-1}$
Column height
Flow rate
Back-pressure
Buffer type and allowed variations in pH and conductivity
Feedstock and buffer storage stability
Acceptable materials in contact with the liquid
Protein sensitivity to shear forces
Recovery based on a well defined product quality assay
Washing procedures
Gel life length
Buffer consumption
Product volumes
Number of product fractions to be collected
Monitoring parameters

to the C-terminus of urogastrone to facilitate its purification and assay. β-Galactosidase,[3,4] chloramphenicol acetyltransferase,[5] and protein A[6] have been used as N-terminal protein fusions. This elegant method combined with a selective affinity or ion exchange step may reduce substantially the cost of purifying proteins and peptides in the future.

Another way to maximize yield at production scale is to ensure recovery of biological activity by using suitable media, equipment and reagents. Temperature, contact time, pH solvents, buffer composition, modifiers and protein concentration also affect protein recoveries.[7] For example, it may be necessary to run the entire separation process at 4°C to preserve biological activity. Sample storage during chromatographic runs and during 'downtimes' must be considered. This may require cold room facilities, which in turn may involve capital investment and maintenance. Running chromatographic processes in the cold may have the unforseen effect of reducing the equilibrium kinetics between the solute and chromatographic media, which may lengthen the process time, resulting in increased labor and utility costs.

Table 27. Costs associated with process chromatography.

Chemicals
Chromatography and auxiliary equipment
Quality control
Sample preparation
Labor
Maintenance
Electricity
Water
Overhead
Facility

It is obvious that costs should be minimized wherever possible. Costs can be broken down as shown in *Table 27*. Chemical costs incurred in chromatography include media, buffers, cleaning agents and high quality water, e.g. water for injection. Solvent recovery systems should be considered when reversed phase HPLC is employed for peptide purification because the expensive (and often toxic) solvents used add significantly to the total cost. One study showed that reducing the cost of solvent from 2.10 to 1.00 US $ per liter reduced operating costs from 87.82 to 69.68 US $ per gram of product.[8] In addition, disposal of hazardous waste, whether solvent or viral-contaminated media, can be quite expensive. In the USA, for example, it costs up to 500 US $ to dispose of a 55 gallon drum of hazardous waste.

Quality control of raw materials, in-process monitoring and final product quality control are part of the costs. An economic analysis of sample preparation, i.e. primary isolation steps, has been carried out for the production of intracellular β-galactosidase in *Escherichia coli*.[9] Labor, overhead and facilities costs should also be considered. Fixed overhead in the USA is typically 500–600 US $ per square meter per year. In production, labor costs tend to run far ahead of supplies and amortized equipment.[10] Automation can minimize labor costs, reduce the bioburden that results primarily from human contact and increase productivity.

The chromatography media cost, which is considered under the chemicals category, may represent a rather large initial investment, but if the media can be used for multiple cycles, the cost per kilogram of product is greatly reduced. Conscientious media maintenance and hygiene (see Chapter 9) can lead to columns with hundreds of cycles.

One economic study of large scale affinity chromatography for the purification of fibronectin from human plasma illustrates how extending the number of chromatographic cycles allows for a cost

Table 28. The purification of fibronectin from human plasma: an estimated costing of chromatographic materials for laboratory scale work (0.5 l plasma).

	US $
Sepharose® 4B (1 l)	169
CNBr-activated Sepharose® 4B (4×15 g)	576
Gelatin (0.2 g, 100 g package)	*
Arginine (500 g)	31
Column K 100/45	1600
Column K 50/30	355
Column K 26/40	185
	2916

*Less than 1.00 US $.

Calculations based on yield of 77 mg, determined from an average of 10 runs.

Table 29. The purification of fibronectin from human plasma: an estimated costing of chromatography material for pilot scale work (5.0 l plasma).

	US $
Sepharose 4B (3 l)	507
CNBr-activated Sepharose 4B (3×250 g)	5925
Gelatin (2 g, 100 g package)	*
Arginine (5 kg)	283
Column K 100/45 (3)	4800
	11 515

*Less than 1 US $.
Calculations based on yield of fibronectin of 770 mg per run extrapolated from laboratory scale.

effective process.[11] Fibronectin, a cell adhesion factor used in tissue culture, was purified using Gelatin-Sepharose® 4B, Arginine-Sepharose 4B and underivatized Sepharose 4B, the latter to eliminate plasma proteins that bind non-specifically to the Sepharose.

Table 28 shows the cost of purifying fibronectin at a laboratory scale from 500 ml of plasma, using a 1 l column of Sepharose 4B, a 150 ml column of Gelatin-Sepharose 4B and a 50 ml column of Arginine--Sepharose 4B. The total cost was 2916 US $ (1986 cost level); the yield per run was 77 mg (average 10 runs).

The investment required for chromatographic media and columns for a pilot scale separation in which fibronectin was purified from 5 l of plasma is shown in *Table 29.* The costs for the purification from 50 l of plasma are shown in *Table 30. Table 31* compares the cost of the materials per milligram of fibronectin produced for (a) laboratory scale, (b) pilot scale and (c) production scale. Assuming the life span of Gelatin-Sepharose 4B is 10 cycles and the life of Arginine-Sepharose 4B is five cycles, it can be seen that the cost of producing fibronectin decreases at the laboratory scale from 86% of the price of fibronectin after one cycle to 2.6% after 100 cycles. (This study assumes that the value of fibronectin is 44.00 US $ mg^{-1}.) At the pilot scale, the cost of chromatography material (media and

Table 30. The purification of fibronectin from human plasma: an estimated costing of chromatography materials for process scale (50.0 l plasma).

	US $
Sepharose 4B (30 l)	4056
Gelatin-Sepharose 4B (commercial product)	18 705
Arginine-Sepharose 4B (commercial product)	6600
Column KS 370A (1)	3950
Column KS 370B (3)	8850
	42 161

Calculations based on yield of fibronectin of 7.7 g per run extrapolated from laboratory scale.

Table 31. The estimated cost of chromatography for each mg fibronectin at laboratory, pilot and production scales.

Number of cycles	Cost per mg (US $)	Percent of price of fibronectin
(a) Laboratory scale:		
1	38.00	86
10	3.92	9
100	1.22	2.6
(b) Pilot scale:		
1	13.00	34
10	1.68	3.8
100	0.62	1.4
(c) Large scale:		
1	5.48	12.4
10	0.63	1.4
100	0.43	0.99

columns) decreases from 34% of the cost of fibronectin after the first cycle to 3.8% after 10 runs and to 1.4% after 100 runs. For the large scale separation (50 l starting material) the cost of the chromatography materials per milligram of fibronectin is 12.4% after one cycle and only 0.99% after 100 cycles.

In this study, the assumption was made that only 10 cycles could be obtained with the Arginine-Sepharose 4B and five cycles with the Gelatin-Sepharose 4B. With careful optimizing and cleaning protocols, the life of these media could be extended (see below). In addition, for most commercial applications, the use of proteinaceous ligands such as gelatin can be avoided. Still, the study shows that the cost of the chromatography media as a percentage of the price of final product is considerably reduced by extending the number of production cycles.

One of the first applications of large scale chromatography for the production of parenteral products was the separation of albumin, IgG and Factor 1X from plasma. *Figure 35* shows the time required and the steps used to obtain albumin and IgG. Factor IX was removed after the DEAE-Sephadex A-50 step and required further purification. Plasma was processed at a rate of 50 l per day and the columns were repeatedly recycled. The variable costs for the production of albumin are shown in *Table 32*. The cost of solvents and buffer salts, as well as separation media, electricity, filter replacement and labor are considered as variable costs. In this example, 10 000 l of plasma were processed per year in 50 l batches and cycled as shown in *Figure 35*.

Table 33 shows the total variable cost added to the process when IgG was also obtained with purity suitable for intravenous use. This cost was 261.88 US $ per 100 l of plasma.

The life of the chromatographic media is given in *Table 34*. (Note that in this process the Arginine-Sepharose 4B lasts 600 cycles—one

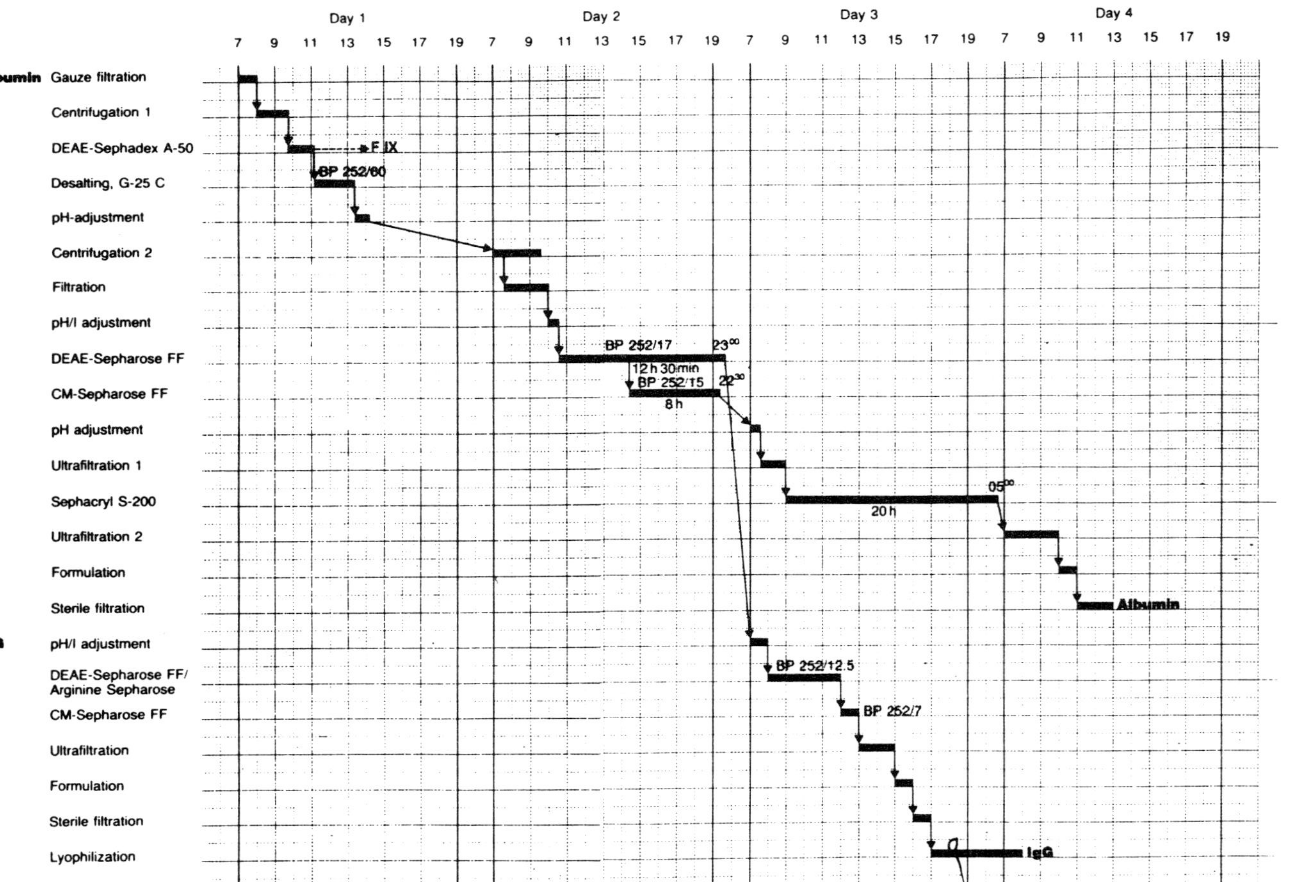

Figure 35. A product schedule for albumin, IgG and Factor IX, with a processing capacity of 50 l plasma day^{-1}.

Table 32. Variable estimated cost calculation: albumin.

	Quantity per 100 l	Price* (US $)	Cost* per 100 l (US $)
NaOH	22.0 kg	1.00 kg^{-1}	22.00
HAc, 60%	61.7 l	0.56 l^{-1}	34.75
NaCl	1.8 kg	0.31 kg^{-1}	0.63
Water	5.0 m^3	7.5 m^{-3}	37.50
NaAc	0.5 kg	2.50 kg^{-1}	1.25
Sephadex G-25 C	40 g	412.50 kg^{-1}	16.50
DEAE-Sepharose FF	200 ml	250.00 l^{-1}	50.00
CM-Sepharose FF	100 ml	250.00 l^{-1}	25.00
Sephacryl S-200	600 ml	112.50 l^{-1}	67.50
Filters†	—	—	143.75
Electricity	230 kWh	0.03 kWh^{-1}	6.38
Labor	4 men @ 12 500 US $ y^{-1}	10 000 l^{-1} y^{-1}	500.00
			905.26

*Purchases and work carried out in Sweden. Cost obtained by using 8 Swedish Kr per US $.
†Assorted micro- and ultra-filters.

Table 33. Variable estimated cost calculation: IgG.

	Quantity per 100 l	Price* (US $)	Cost* per 100 l (US $)
NaOH	2.6 kg	1.00 kg^{-1}	2.63
HAc, 60%	0.6 l	0.56 l^{-1}	0.38
NaCl	0.5 kg	0.31 kg^{-1}	0.12
Glycine	0.4 kg	2.50 kg^{-1}	1.00
Water	1.0 m^{-3}	7.50 m^{-3}	7.50
DEAE-Sepharose FF	50 ml	250.00 l^{-1}	12.50
CM-Sepharose FF	20.0 ml	250.00 l^{-1}	5.00
Arginine-Sepharose	30.0 ml	1250.00 l^{-1}	37.50
Filters	—	—	64.25
Electricity	220.0 kWh	0.28 kWh^{-1}	6.00
Labor	1 man	—	125.00
			261.88

*Purchases and work carried out in Sweden. Cost obtained using 8 Swedish Kr per US $.

Table 34. The lifetime of chromatographic media for albumin/IgG process.

	Lifetime (cycles)	Lifetime (years)
Albumin		
Sephadex G-25C	7200	2
DEAE-Sepharose FF	600	0.5
CM-Sepharose FF	1200	1
Sephacryl S-200	800	1
IgG		
DEAE-Sepharose FF	600	1
Arginine-Sepharose	600	1
CM-Sepharose FF	400	2

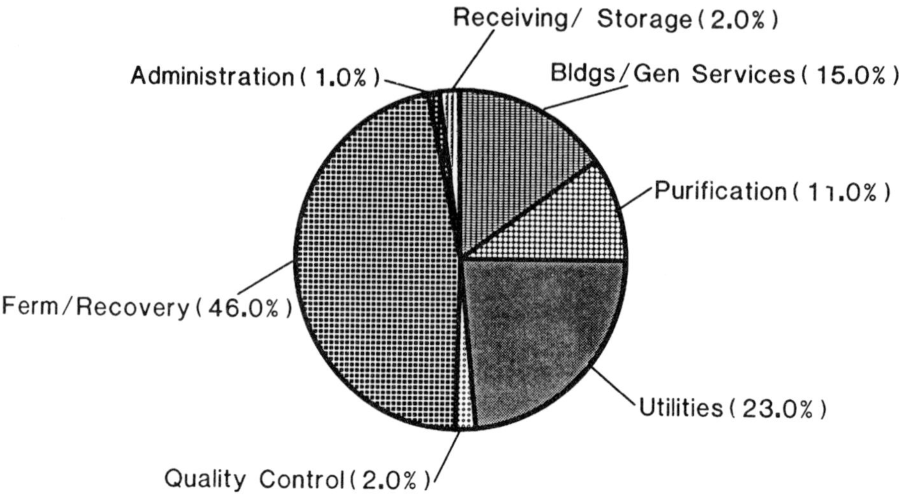

Figure 36. A breakdown of plant investment. (Reproduction by kind permission of Arthur D. Little, Inc.)

Table 35. Yield, estimated cost and estimated market value of albumin and IgG.

Albumin	
Yield per 100 l plasma (kg)	2.5
Cost per kg albumin (US $)	362.50
Market value per kg albumin (US $)	3125.00
IgG	
Yield per 100 l of plasma (kg)	0.5
Cost per kg IgG (US $)	525.00
Market value per kg IgG (i.v.) (US $)	49 375.00

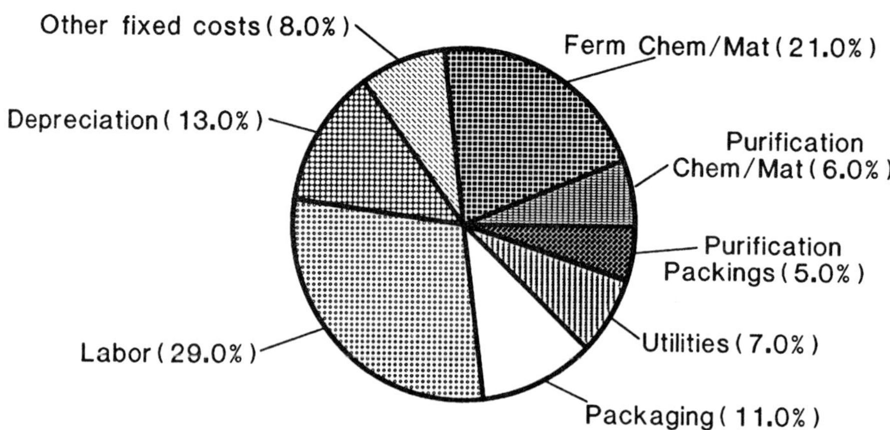

Figure 37. A breakdown of manufacturing costs. (Reproduction by kind permission of Arthur D. Little, Inc.)

year—as compared to an estimated 10 cycles shown in the fibronectin application. Since plasma fractionation plants are run in several locations around the world, this number reflects a more accurate calculation.) The life of the DEAE-Sepharose® Fast Flow is half that of the CM-Sepharose Fast Flow. This is because the DEAE step is first in the sequence and the sample has more contaminants. Since the first separation medium used generally requires more frequent replacement, it is advisable to use more expensive media at a later step in the purification scheme.

Table 35 shows the yield of albumin (2.5 kg) and IgG (0.5 kg) from 100 l of plasma. Examination of *Table 32* indicates that the total variable costs were 905.26 US $ for the production of albumin from 100 l of plasma. Since 2.5 kg of albumin were obtained from 100 l of plasma, the cost per kilogram of albumin was 362.50 US $. The market value per kilogram of albumin is 3125.00 US $ (1985 price level). Thus the variable costs were only 11.6% of the market value. The same calculation for IgG shows that the variable costs were only 1% of the market value.

Today, many engineering and architectural firms are actively involved in the design and construction of facilities for pilot and production plants. A review of important criteria and average costs per gross area unit for construction, process/piping systems, mechanical HVAC systems and electrical systems has been published.[12] Breakdown of plant investment and manufacturing costs are shown in *Figures 36* and *37*. It is interesting to note that the plant investment for purification is 11% of the total, whereas fermentation and recovery make up 46% of the total investment. For manufacturing, purification packings (media) are estimated to be 5% of the total cost. After adding costs for other chemicals and materials, total purification costs come to only 11% of the total manufacturing costs.

In summary, there are a number of key factors influencing the cost-effectiveness of a chromatographic process. These include: reducing the number of steps, automating to lower labor costs and increase productivity, and lowering variable production costs by increasing media life. These, as well as several other factors discussed, should be considered at the early stages of process design to maximize profitability.

REFERENCES

1. Johansson, H., Lindquist, L.-O. and Low, D. Design criteria for the purification of biopolymers for the pharmaceutical industry. Eurochem-Process Engineering Today, Birmingham, England, 3-5 June 1986.

2. Kamen, R. Production of heterologous human proteins by fermentation. First ASM Annual Conference on Biotechnology, Washington, DC, 20-23 March 1986.
3. Ullmann, A. 1-Step purification of hybrid proteins which have beta galactosidase activity. *Gene* **29** (1984) 27-32.
4. Germino, J. and Bastia, D. Rapid purification of a cloned gene product by genetic fusion and site-specific proteolysis. *Proc. Natl Acad. Sci.* **81** (1984) 4692-4696.
5. Bennet, A., Rhind, S. K. and Lowe, P. International patent publication number WO 84/04756.
6. Nilsson, B., Holmgren, E. and Josephson, S. Efficient secretion and purification of human insulin-like growth factor I with a gene fusion vector in *Staphylococci. Nucleic Acids Res.* **13** (1985) 1151-1162.
7. Johnson, R. D. Preparative high performance liquid chromatography of peptides and proteins. Developments in Industrial Microbiology. *J. Ind. Microbiol. Suppl. No. 1*, **27** (1987) 77-83.
8. Large-scale HPLC for biotech: a reality! *Biotechnology News* **4** (1984) 6.
9. Datar, R. Economics of primary separation steps in relation to fermentation and genetic engineering. *Process Biochem.* **21** (1986) 19-26.
10. Wheelwright, S. M. Designing downstream processes for large-scale protein purification. *Bio/Technol.* **5** (1987) 789-793.
11. Janson, J.-C. Scaling-up of affinity chromatography: technological and economical aspects. In *Affinity Chromatography and Related Techniques* (Gribnau, T. C. J., Visser, J. and Nivard, R. J. F., eds). Elsevier, Amsterdam, 1982, pp. 503-512.
12. Kennedy, R. N. and Thompson, R. G. R&D facility budgets, and costs that drive design. *Pharm. Eng.* **7** (1987) 13-17.

Appendix A Chromatographic Techniques: An Overview

In any chromatographic separation, the goal is to resolve two or more components. Resolution is a function of efficiency, selectivity, and capacity:

$$R_s = 1/4 \sqrt{N} \, \frac{\alpha - 1}{\alpha} \, \frac{K' + 1}{K} \tag{2}$$

where α=selectivity, K' =capacity and N=efficiency[2]. As shown in *Figure 38*, each of these terms is defined by an equation which contains directly measurable parameters. Chromatographic theory, derived from size exclusion chromatography, describes selectivity as a measure of the distance between peak maxima. Selectivity can be determined by measuring the elution volume V_{e1} of the first chromatographic peak and the elution volume V_{e2} of the second peak, as well as the void volume V_0 (the elution volume of a molecule too large to enter the pores of the chromatographic gel). Selectivity can be enhanced by selecting the proper media. Particle size distribution, ligand and solvent systems also influence selectivity.

Efficiency is measured by determining the elution volume V_e and the width W of the chromatographic peak. Efficiency is a function of the way in which the column is packed and is dependent on media particle size, column design and system optimization. According to Janson and Hedman,[1] it is possible to calculate the production rate P (i.e. the quantity of protein product that can be separated per unit time) for different values of the column plate number, N, which is a measure of column efficiency. It is clear from *Figure 39*, that "above a certain level very little is gained in production rate by increasing the column efficiency, i.e. increasing the plate number, N, by, for example, reducing the particle size."

Capacity, readily measured as shown in *Figure 38*, is a measure of solute retention; it can be increased by changing the characteristics of the mobile phase. The dynamic capacity (i.e. the binding capacity

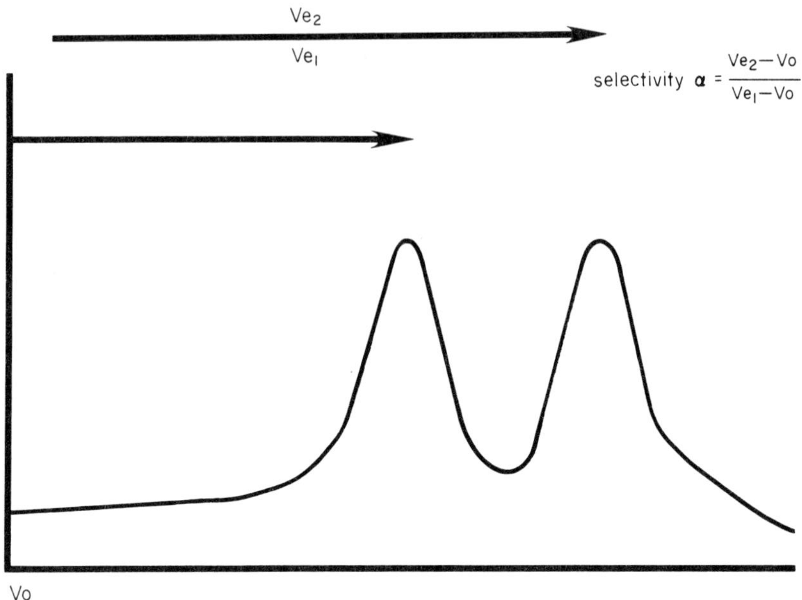

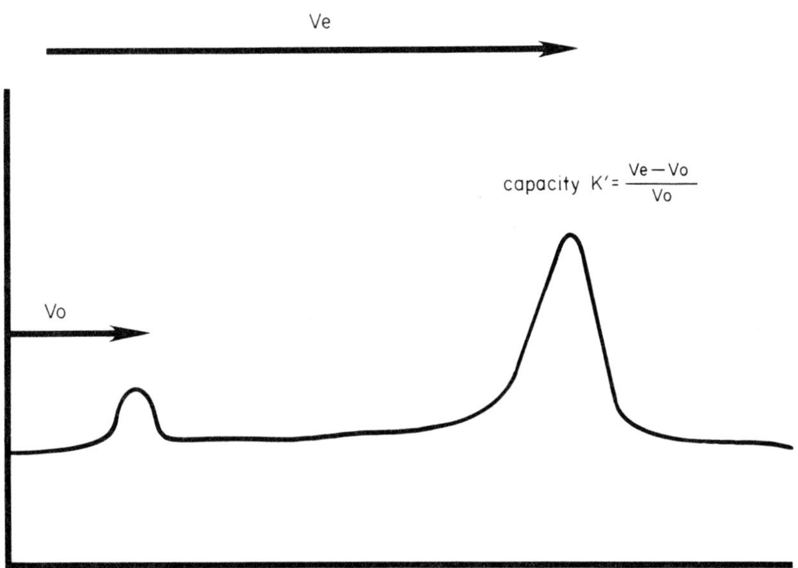

Figure 38. (See page 119 for caption)

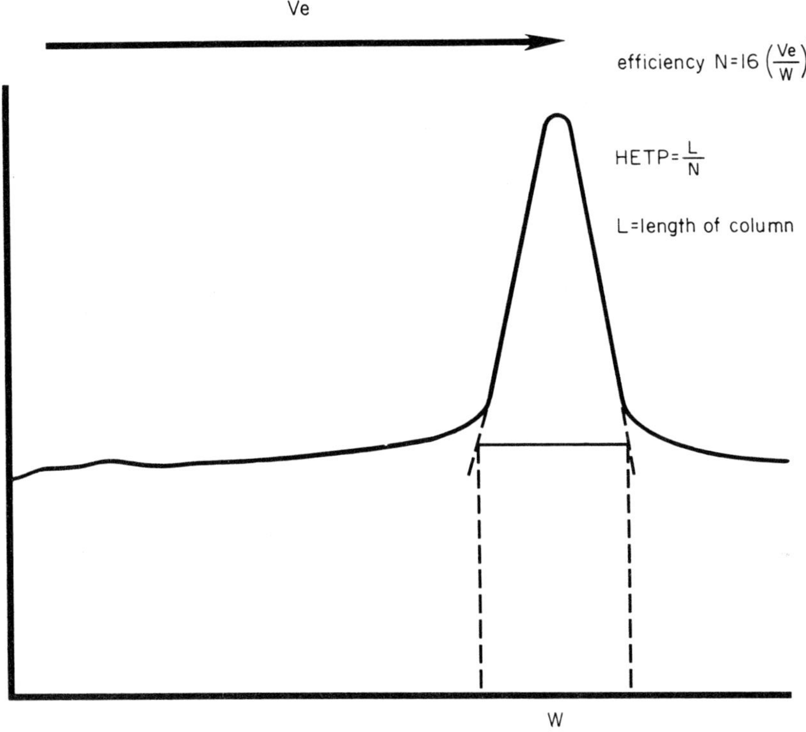

Figure 38. Selectivity, capacity and efficiency are directly measured parameters.

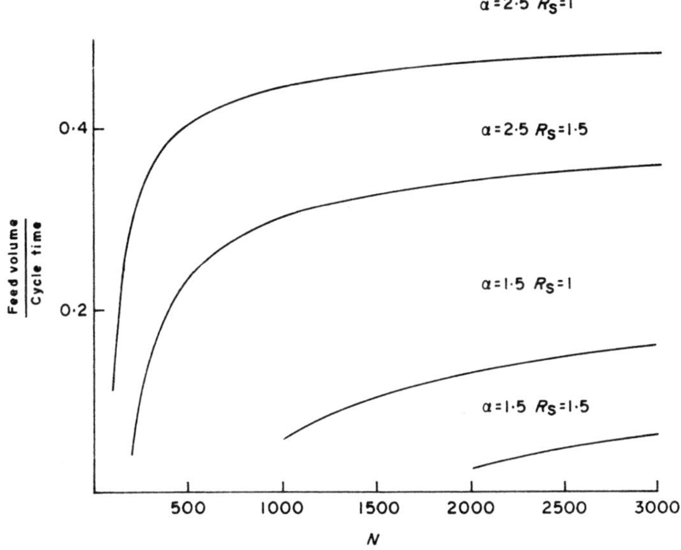

Figure 39. Dependence of production rate on column plate number (*N*) for different combinations of the selectivity factor, α, and resolution, R_s.

of chromatography adsorbents under working conditions) of the column packing material is one of the key factors in determining throughput. And throughput is one of the most important factors for industrial scale chromatography. According to Janson and Hedman, the dynamic capacity for a particular solute is dependent on several factors: matrix composition and pore structure, particle diameter and particle size distribution profile, solute molecular weight and solubility, forward and backward rate constants of the binding reaction, bulk-, film- and gel-diffusion constants of the solute and, finally, possible competitive binding and displacement effects of other proteins present in the sample feed solution.

For further details on chromatography theory, the reader is referred to Ref. 2.

There are two basic kinds of chromatographic procedures: adsorption and partition. Adsorption techniques have a higher capacity than partition techniques and should, therefore, be the first choice, especially in the initial stages of a purification scheme where there is more volume to process and more contaminants to remove. Chromatographic adsorption techniques include: ion exchange, hydrophobic interaction and affinity chromatography.

GEL FILTRATION

The major partition technique is gel filtration (also known as size exclusion chromatography or gel permeation chromatography). In gel filtration, molecules are partitioned between the stationary phase (separation medium) and the mobile phase (solvent system) (see *Figure 40*). Large molecules which cannot penetrate into the pores of the stationary phase are excluded from the liquid inside the separation medium and are eluted first. Intermediate size molecules diffuse into a limited proportion of the stationary phase, while small molecules diffuse freely into the stationary phase.

A selectivity curve can be used to determine the optimal separation medium for a particular application (*Figure 41*). The greater the slope in the linear portion of the curve (*Figure 41a*), the greater will be the difference in elution volumes of two molecules. As a guideline, the product should elute with a K_{av} of between 0.1 and 0.5. This will ensure that the product does not elute in the void volume and that band broadening does not occur due to dilution.

There are two main uses for gel filtration as a separation tool—group separation and fractionation. In group separation, the objective is to separate a high molecular weight substance, e.g. a protein, from low molecular weight substances, e.g. peptides, growth media or salts (desalting). Group separation can also be used for buffer exchange. Flow rates are high and sample volume is typically 20–30% of the total gel bed volume. The columns used for group fractionation

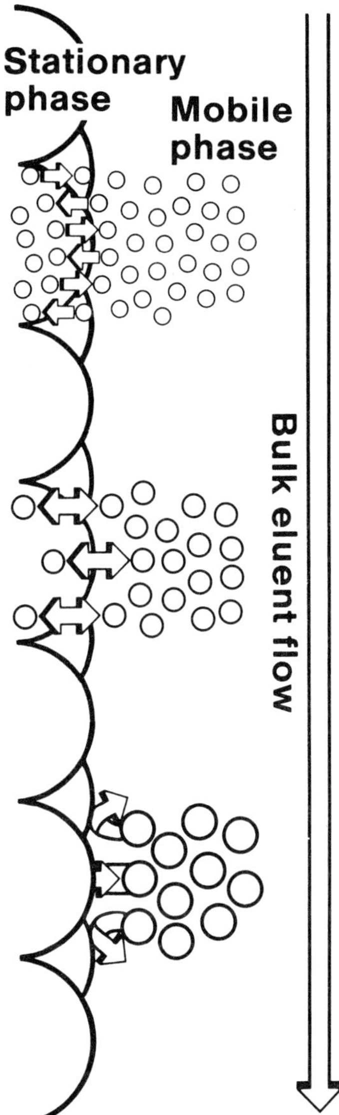

Figure 40. The mechanism of gel filtration.

are short and wide. In contrast, when fractionating molecules that are close in size, the sample volume is limited to anywhere from 2 to 10% of the total gel bed volume, and lower flow rates are usually required to achieve the desired resolution. Longer columns must also be used. In spite of these limitations, fractionation by gel filtration is often an excellent final chromatographic step. It can be used, for example, to separate aggregates which occur during the process, and at the same time it can exchange the buffer the product is in so that it is ready for final packaging.[3]

(a)

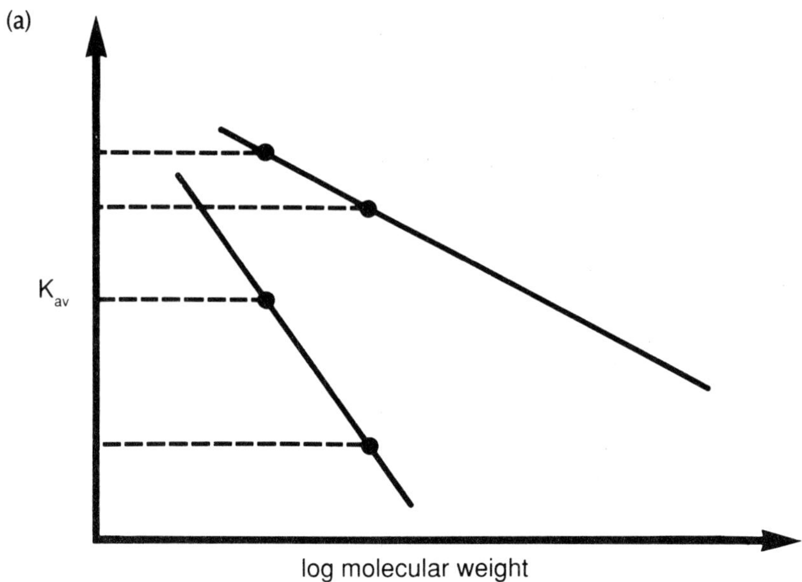

$$K_{av} = \frac{V_e - V_o}{V_t - V_o}$$

where V_e = elution volume
V_o = void volume
V_t = total bed volume

(b)

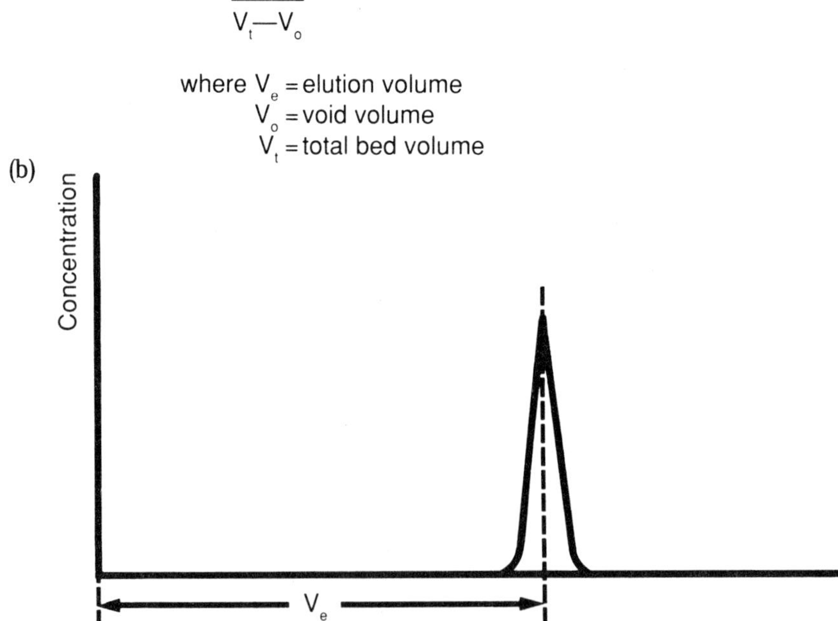

Figure 41. Selectivity curves are used to determine the optimal gel filtration medium. (a) Selectivity curve. (b) Measurement of K_{av}.

Table 36. Biospecific complexes.

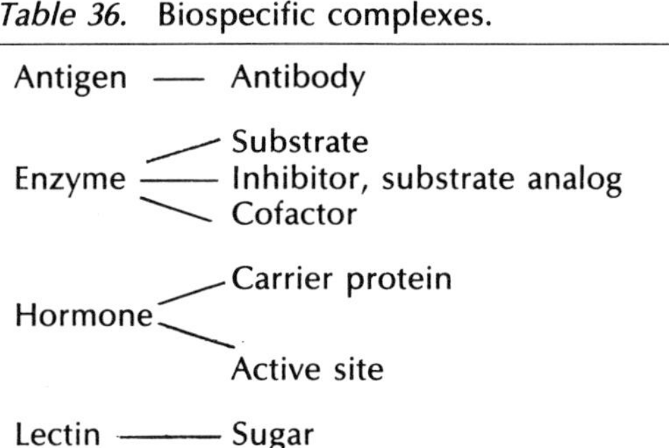

Antigen	—	Antibody
Enzyme	—	Substrate
		Inhibitor, substrate analog
		Cofactor
Hormone	—	Carrier protein
		Active site
Lectin	—	Sugar

ION EXCHANGE

In ion exchange chromatography, proteins and peptides are separated due to their charge differences. This technique is well suited for large scale applications because it has a high capacity (dynamic capacity typically 30 g total protein per liter of adsorbent),[4] is selective, has a concentrating effect, and can be used to isolate many proteins in one step. Flow rates are high, and short, wide columns are used (see *Table 36*). Throughputs of 4.5 kg can be cycled in less than 1 h.

Figure 42 shows some ion exchanger types. It is important to select an ion exchanger that can be used in the pH range in which the molecule of interest is stable. In selecting the ion exchanger, the following should be considered: capacity for the molecule(s) of interest, ability of the matrix and functional group to withstand high flow rates and the ability of the separation media to be cleaned in place. Usually, after the sample is adsorbed, the components of interest are eluted from the ion exchanger with continuous or step gradients of pH and/or increasing ionic strength. Sometimes, ion exchangers are used to bind the contaminants and the molecule(s) of interest pass through the column. In this case, the concentrating effect is not achieved.

AFFINITY CHROMATOGRAPHY

Affinity chromatography is an exquisitely selective technique that takes advantage of the biospecific interactions that occur in nature (see *Table 36*). Purifications up to 2×10^6-fold have been achieved in a single step (see *Figure 43*). Because the desired degree of purity can sometimes be achieved in a single step, affinity chromatography is an excellent purification tool for biotechnology. Recent reviews on affinity chromatography are indicative of the high level of interest in this technique.[5-7] *Table 37* shows some large-scale affinity chromatography applications. Flow rates used in affinity chromatography are

Designation	Group	Counter Ion
DEAE	$-O-C_2H_4-\overset{+}{N}H(C_2H_5)_2$	Cl^-
QAE	$-O-C_2H_4-\overset{+}{N}(C_2H_5)_2-CH_2-CH(OH)-CH_3$	Cl^-
CM	$-O-CH_2-COO^-$	Na^+
SP	$-O-C_3H_6-SO_3^-$	Na^+

Ion Exchanger	Recommended pH Range
DEAE	2–9
QAE	2–10
CM	6–10
SP	2–10

Figure 42. Types of ion exchanger.

high, columns are short and wide and, as shown in *Table 38*, it is an easy-to-perform technique.

There are many types of matrices and many chemistries available for coupling ligands to the gel matrix. (For further reading, see Ref. 8.) *Table 39* shows some of the ways in which a matrix can be activated, the group on the ligand that reacts with the activated group, and the resulting coupled separation medium. Cyanogen bromide (CNBr), tresyl, epoxy, *N*-hydroxy succinimide, carbonyldiimidazole, thiol and diazonium activated media are commercially available. Epoxy activated media give very stable bonds and can be used to couple hydroxyl, sulfhydryl and amino groups on the ligand. The pH for coupling is rather high (8–13 for amino groups, 10–13 for sulfhydryl groups and 11–13 for hydroxyl groups) and this can be a problem for very labile ligands. Tresyl activated agarose media are produced by activating the agarose hydroxyl groups with tresyl chloride (2,2,2-trifluorethanesulfonyl chloride). This medium can be swollen and stored for up to 1 month prior to coupling the ligand. Amino, sulfhydryl, or hydroxyl groups can be coupled to tresyl activated media. *N*-Hydroxy succinimide media react with primary amino groups, while the thiol activated media react with free thiol groups and the diazonium gels react with aromatic alcohols. Cyanogen

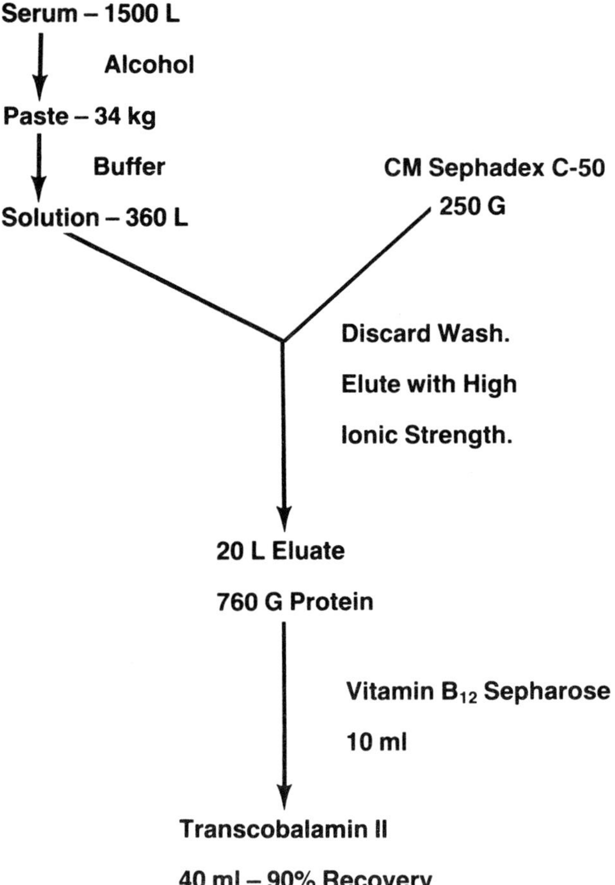

Figure 43. The purification of transcobalamin II from serum. (From Allen, R. H. and Majerus, P. W. *J. Biol. Chem.* **247** (1972) 7709–7717.)

bromide activated media couple the ligand through amino groups.[9] Optimization of activation conditions to provide stable coupling and the physiological coupling conditions make this the most widely used chemistry for affinity chromatography.

One important aspect of affinity chromatography to consider is leakage. In spite of some claims, there is no such thing as leak-proof media. By using sensitive detection methods, leakage can be quantified; toxicity, if any, of the leakage product can be determined. When the ligand is attached through multiple sites, less leakage occurs. At a recent symposium,[10] one vaccine manufacturer stated that alternatives to CNBr coupling of antibodies have not offered any more stable coupling. Today, there are several products with FDA approval that are purified by affinity chromatography. In one case, a non-toxic, small molecular weight ligand is coupled via a single point attachment by CNBr activated Sepharose®.

Table 37. Large scale affinity chromatography applications.

Protein	Adsorbent
Plasminogen	Lysine-Sepharose 4B
Antithrombin III	Heparin-Sepharose CL-6B
α_1-antitrypsin	Activated Thiol-Sepharose 4B ConA-Sepharose
Albumin	Blue Sepharose CL-6B
Human serum amylase	Cycloheptaamylose coupled to Epoxy activated Sepharose 6B
Transcortin	Cortisol-agarose
Angiotensinogen	ConA-Sepharose
Human fibroblast interferon	ConA-Sepharose Phenyl-Sepharose CL-4B
Glucocerebroside-β-glucosidase	Decyl-agarose Octyl-agarose
Hog Kidney Renin	Pepstatin coupled to AH-Sepharose 4B
3-Hydroxybutyrate dehydrogenase	Procion blue Sepharose
Glycerokinase	Procion blue Sepharose
β-Galactosidase	p-Aminophenyl-β-C-galactopyranoside coupled Sepharose 4B
Kunitz protease inhibitor	Immobilized trypsin
Elastase	Trasylol coupled Sepharose
Crotalaria juncea lectin	Galactose coupled Sepharose 6B
Tetanus toxoid antibodies	Spherosil-DEAE-Dextran-toxoid antibody-toxoid
Lactate dehydrogenase	NAD-coupled AH-Sepharose 4B
Malate dehydrogenase	NAD-coupled AH-Sepharose 4B
Glutamate-oxaloacetate transaminase	Pyridoxal-5-phosphate coupled AH-Sepharose 4B
Glutamate-pyruvate transaminase	Pyridoxal-5-phosphate coupled AH-Sepharose 4B

Table 38. Critical stages in affinity chromatography.

Choice of matrix
Choice of ligand
Coupling
Sample application
Adsorption of sample
Wash away unbound substances
Elute bound substances
Desalt
Regenerate gel

Table 39. Activation chemistries for affinity chromatography.

Activation chemistry	Reaction
Cyanogen bromide	Ⓜ—O—C≡N + RNH_2 → Ⓜ—O—C(=NH)—NHR
Tresyl chloride	Ⓜ—CH_2—O—SO_2—CH_2—CF_3 + RNH_2(RSH) → M —CH_2—NHI
Epoxy	Ⓜ—CH(—O—)CH_2 + RNH_2(ROH,RSH) → Ⓜ—CH(OH)—CH_2—NH—R
N-hydroxysuccinimide	Ⓜ—C(=O)—O—N(succinimide) + RNH_2 → Ⓜ—C(=O)—NH—R
Carbonyldiimidazole	Ⓜ—O—C(=O)—N(imidazole) + RNH_2 → Ⓜ—O—C(=O)—NHR
Thiol	Ⓜ—S—S—(2-pyridyl) + RSH_2 → Ⓜ—S—S—R
Diazonium	Ⓜ—C_6H_4—N≡N^+ + R—C_6H_4—OH → Ⓜ—C_6H_4—N≡N—C_6H_3(HO)(R)

Since the development of hybridoma technology, there has been an increased interest in the use of immunoadsorbents in which a monoclonal antibody is coupled to a matrix. For example, human interferon produced in *Escherichia coli* has been purified in a semi-continuous process using tangential-flow microfiltration and a monoclonal antibody affinity column.[11] Antigen binding capacity of immunoadsorbents has been optimized.[12] Immunoadsorbents are probably the most specific affinity media.

Group specific affinity adsorbents, while not as selective, can reduce the media cost. Dye-ligand chromatography has been used for many large scale purifications. An extensive list of applications is given by Qadri.[13] The dyes are inexpensive and relatively stable; their protein binding capacity is high. In the large scale purification of glycerokinase from *Bacillus stearothermophilus*, a final step of dye affinity chromatography replaced three steps in the conventional, small scale purification and gave excellent recovery of homogeneous enzyme.[14]

Another widely used group specific affinity technique suitable for process scale is immobilized metal affinity chromatography (IMAC). This technique is quite useful for the purification of plasminogen activators[15,16] and interferons.[17] A recent review describes mechanisms involved in IMAC.[18]

HYDROPHOBIC INTERACTION CHROMATOGRAPHY

Hydrophobic interaction chromatography (HIC) is another adsorption technique. This technique takes advantage of the existence of hydrophobic 'pockets' on the surface of many proteins and peptides. Hydrophobic interaction media are substituted with hydrophobic functional groups—alkyl groups, e.g. octyl groups (C_8) or aryl groups, e.g. phenyl. Generally, the affinity of proteins for the separation media surface increases as the chain length of the ligand increases.[19] The driving force of the interaction between the protein and the functional group is an increase in entropy.

HIC is quite useful in the early stages of purification since it is compatible with high salt, e.g. ammonium sulfate, that may be used in an initial precipitation step. High ionic strengths, salts which have a high degree of 'salting-out' ability, and decreased pH promote the binding of hydrophilic as well as hydrophobic proteins. It is often possible to precipitate the bulk of the contaminants and bind the components of interest at a salt concentration just below that which causes them to precipitate. Elution can be achieved by lowering salt concentrations, by decreasing polarity with polarity reducing agents such as ethylene glycol or by using deforming buffers which cause conformational changes in the protein. Either linear or step gradients can be used.

HIC has a concentrating effect, and since the mobile phases used are generally compatible with most proteins, biological activities can be preserved.

REVERSED PHASE CHROMATOGRAPHY

Reversed phase chromatography (RPC) also separates molecules according to their different hydrophobicities. This technique, however, employs less hydrophilic support matrices, and the supports are more highly substituted. RPC media are essentially of two types: silica supports with a hydrocarbon bonded phase (usually straight chain alkyl groups, covalently linked to the support) and non-silica polymers in which the separation medium is both the physical support and the non-polar phase. Historically, RPC has been widely used for the purification of small molecules, such as synthetic drugs, pesticides and metabolites. Although the basic retention mechanism is probably the same in HIC and RPC, native protein structure is more likely to be maintained in HIC.[20] In the production of recombinant proteins from, e.g. *E. coli*, the bacteria often isolates the 'foreign' protein as an inclusion body. When these inclusion bodies are isolated, the protein is denatured and must be extracted with solvents such as acetonitrile or urea. In this case, RPC is useful because of its high capacity and excellent resolving capabilities.

Table 40. The influence of hydrophobicity on the selection of separation technique and eluting solvent.

	Increasing hydrophobicity →			
	Native			Denatured
	Ion exchange	Pseudo-hydrophobic interaction	Hydrophobic interaction	Reversed phase
Separation property	Charge	Charge Hydrophobicity	Hydrophobicity	Hydrophobicity
Typical group	DEAE CM	Aminohexyl Carboxyhexyl	Phenyl Octyl	C_8 C_2
Typical eluting solvent	Water	Water Ethylene glycol	Water Ethylene glycol	Methanol Ethanol Acetonitrile

In RPC, the strong interaction of the peptide or protein with the highly substituted matrix requires the use of organic solvents for elution (see *Table 40.*) Acetonitrile, 2-propanol and methanol are the solvents most frequently used for elution. 2-Propanol, which has the greatest eluting strength, is best used with strongly adsorbed hydrophobic peptides. Methanol is best with weakly adsorbed hydrophilic peptides. Both 2-propanol and methanol give high back-pressures. Methanol has a high heat of mixing which tends to cause solvent degassing during a run. Acetonitrile is the preferred solvent for most applications. It combines good eluting strength with low back-pressure. The favorable mass transfer properties of acetonitrile also lead to good column efficiencies.

Isocratic elution is frequently used for elution of small molecules in RPC, but for peptides, gradient elution provides the best resolution. Generally, the two mobile phase components are designated as mobile phase A (relatively polar) and mobile phase B (more non-polar than A).

The most common mobile phase A components are aqueous acids at low pH. The low pH (less than 3) results in ion suppression of the silanol groups on silica columns. This results in sharp peaks with no trailing. The most popular acids are hexafluorobutyric acid (HFBA), trifluoroacetic acid (TFA) and phosphoric acid. Retention times for peptides are generally greatest for HFBA and least with phosphoric acid. TFA, which tends to ion pair to the peptide, is most commonly used. An additional advantage of TFA is that it is volatile so that the solvent can be removed from the peptide by lyophilization. Use of phosphoric acid, on the other hand, requires a desalting step after the RPC column. Detection in the presence of some mobile phases can be a problem. Both TFA and HFBA can cause drifting baselines since they absorb at 214 nm.

REFERENCES

1. Janson, J.-C. and Hedman, P. On the optimization of process chromatography of proteins. *Biotechnol. Prog.* **3** (1987) 9-13.
2. *Introduction to Modern Liquid Chromatography* (Snyder, L. R. and Kirkland, J. J., eds.). Wiley Interscience, New York, 1979, 2nd Edn.
3. Note. Flow rates in chromatography are often expressed cm h^{-1}. This is the same as (ml cm^{-2} h^{-1}). Volumetric flow rates (ml h^{-1}) are calculated according to the formula: π (cross-sectional area of the column)$^2 \times$flow rate.
4. Scopes, R. *Protein Purification, Principles and Practice.* Springer-Verlag, New York, 1982.
5. Parikh, I. and Cuatrecasas, P. Affinity chromatography. *Chem. Eng. News* **63** (1985) 17-32.
6. Walters, R. Affinity chromatography. *Anal. Chem.* **57** (1985) 1099-1114.
7. Clonis, Y. S. Large-scale affinity chromatography. *Bio/technol.* **5** (1987) 1290-1293.
8. *Affinity Chromatography, A Practical Approach* (Dean, P. D. G., Johnson, W. S. and Middle, F. A., eds). IRL Press, Washington, DC, 1985.
9. Axen, R., Porath, J. and Ernback, S. Chemical coupling of peptides and proteins to polysaccharides by means of cyanogen halides. *Nature* **214** (1967) 1302-1304.
10. Wampler, D. E. Hepatitis B vaccine purification by immunoaffinity chromatography. In *Modern Approaches to Vaccines: Molecular and Chemical Basis of Virus Virulence and Immunogenicity* (Chanock, R. M. and Lerner, R. A., eds) Cold Spring Harbor Laboratory, Cold Spring Harbor, 1983. Vol. XXVII.
11. Vaks, B., Mory, Y. and Pederson, J. U. A semi-continuous process for the production of human interferon from *E. coli* using tangential-flow microfiltration and immuno-affinity chromatography. *Biotechnol. Letts* **6** (1984) 621-626.
12. Pfeiffer, N. E., Wylie, D. E. and Schuster, S. M. Immunoaffinity chromatography utilizing monoclonal antibodies. *J. Immunol. Methods* **97** (1987) 1-9.
13. Qadri, F. The reactive triazine dyes: their usefulness and limitations in protein purifications. *Trends Biotechnol.* **3** (1985) 7-11.
14. Scawen, M. D., Hammond, P. M. and Comer, M. J. The application of triazine dye affinity chromatography to the large-scale purification of glycerokinase from *Bacillus stearothermophilus. Anal. Biochem.* **132** 413-417.
15. Rijken, D. D. and Collen, D. Purification and characterization of the plasminogen activator secreted by human melanoma cells in culture. *J. Biol. Chem.* **256** (1981) 7035-7041.
16. Rijken, D. C., Wijngaards, G., Zaal-de Jong, M., Welbergen, J. *et al.* Purification and partial characterisation of plasminogen activator from human uterine tissue. *Biochim. Biophys. Acta* **580** (1979) 140-153.
17. *Metal Chelate Affinity Chromatography Reference List.* Pharmacia, Sweden, 1984.
18. Sulkowski, E. Purification of proteins by IMAC. *Trends Biotechnol.* **3** (1985) 1-6.
19. Arfmann, H. A. and Shaltiel, S. Resolution and purification of histones on homologous series of hydrocarbon-coated agaroses. *Eur. J. Biochem.* **70** (1976) 269-74.
20. Fausnaugh, J. L., Kennedy, L. A. and Regnier, F. E. Comparison of hydrophobic-interaction and reversed-phase chromatography of proteins. *J. Chromatogr.* **317** (1984) 141-155.

Appendix B *Product Analysis*

The required purity of the product will be dictated by its final use. The specifications will be set during process development. Some therapeutics obtained by recombinant DNA technology are today reportedly purified to 99.999%.[1]

Purity criteria generally fall into two categories: biological and biochemical. In addition to determining that the product has the specified biological activity, immunological data and the absence of antiviral activity, pyrogenic activity and exogenous DNA must be shown. Several analytical techniques are usually performed on the final product. Some of the techniques used to characterize a protein are shown in *Table 41*.[2]

Table 41. Physicochemical characterization of proteins and peptides.

Amino acid composition
DNA content
Carbohydrate analysis
Circular dichroism
FAB mass spectrometry
High performance liquid chromatography
Isoelectric focusing
Lipid analysis
Optical rotary dispersion
Partial sequence analysis
Peptide mapping
Polyacrylamide gel electrophoresis

The characterization of the product must be sufficiently detailed to show that the product is not contaminated by unwanted constituents and that the molecular structure is not altered. For example, techniques used in the analysis of human growth hormone include: biological assay, determination of correct disulfide bond formation by peptide mapping and secondary structure analysis by circular dichroism.[3] Human α-interferon was produced in *Escherichia coli* by recombinant DNA techniques, and the final product was purified to apparent homogeneity from approximately 1000 l of fermentation broth with a yield of 19%. The final product was analyzed for antiviral activity, endotoxin levels and protein concentration. In addition to these assays, reversed phase HPLC and two dimensional gel electrophoresis were used for in-process monitoring and analysis of the final product.[4] For further reading see Refs 5 and 6.

REGULATORY TERMS[7]

IND investigational new drug application: required prior to performance of clinical investigations on human subjects.

NDA new drug application: required to be approved before new drug can be marketed.

ANDA abbreviated new drug application (for products which have already been on the market for a considerable length of time, e.g. generic drugs).

DMF Drug Master File: a reference source providing detailed information about a specific facility, process or article used in the manufacture, processing, packaging or holding of a substance which is the subject of an investigational new drug application, a new drug application, abbreviated new drug application or Antibiotic Form 5 or 6. The confidentiality of DMFs is assured in accordance with regulations of the US Department of Health, Education and Welfare, Public Health Service, Food and Drug Administration.

cGMP current Good Manufacturing Practices: GMP regulations for drugs and biologics are set forth in document 21 CFR 210-211 and in Good Laboratory Practices (21 CFR 3e); GMP regulations for medical devices are in 21 CFR 820.

PMN pre-market notification, commonly called 510(k)—notification for most medical devices.[8]

PMA pre-market approval, for most 'critical' (e.g. life sustaining) devices (equivalent to an NDA).

GRAS generally regarded as safe.

GILSP good industrial large scale producing organism (European Federation of Biotechnology).

PRODUCT LICENSE APPLICATION equivalent to an NDA, but for products defined as 'biological products' by the FDA (e.g. vaccines, antitoxins, blood products, allergenic extracts, immune modulators, etc.).

REFERENCES

1. Personal communication.
2. Gates, F. T., III. Regulating new technologies. *Pharm. Eng.* **7** (1987) 22-25.
3. Becker, G. W. and Hsiung, H. M. Expression, secretion, and folding of human growth hormone in *Escherichia coli. FEBS* **204** (1986) 145-55.
4. Trotta, P. P., Van Le, H., Sharma, B. and Nagabushan, T. L. Isolation and purification of human alpha interferon, a recombinant DNA protein. *J. Ind. Microbiol.* **27** (1987) 53-64.
5. Bogdansky, F. M. Considerations for the quality control of biotechnology products. *Pharm. Technol.* **11** (1987) 72-74.
6. Committee for Proprietary Medicinal Products *Ad Hoc* Working Party on Biotechnology/Pharmacy. Guidelines on the production and quality control of medicinal products derived by recombinant DNA technology. *TIBTECH* **5** (1987) 1-4.
7. For further information, see *Bio-factors* **1**(1) (1986) 14-17.
8. Note. Chemicals that have no toxic effects and are used for their mechanical properties are considered medical devices.

Appendix C Regulatory Considerations

Regulatory considerations are often as critical to the successful production of a biological as the purification scheme. This is true especially for those products produced by modern hybridoma and recombinant DNA technologies.

For process validation, it will be necessary to establish that the product is consistently produced, and not contaminated by unwanted materials. Process validation is a good manufacturing practice (GMP) quality assurance measure intended to protect end-product fitness for use. Validation provides documented evidence that a process does what it purports to do, reliably and consistently. A validated process operates in a state of control and includes provisions for recognition of deviations, so that corrective actions can be taken.[1]

For products produced by recombinant DNA or hybridoma technologies, it will also be necessary to describe the materials and methods used to construct the recombinant gene as well as the final production strain. If the product is produced by cell culture, it will be necessary to assure that the final product is purified so that there is no cellular DNA, host cell antigens or adventitious agents present. Details of fermentation, purification and quality assurance testing must be presented. The issues of containment in fermentor areas, during cell harvesting and initial recovery steps as well as decontamination and waste disposal are discussed by Gates.[2]

The concerns of the US Food and Drug Administration are also expressed in several documents entitled *Points to Consider*.[3]

For further information see Refs 4-19.

REFERENCES

1. PMA's Deionized Water Committee. Validation and control concepts for water treatment systems. *Pharm. Technol.* **9** (1985) 50-56.
2. Gates, F. T. Regulating new technologies. *Pharm. Eng.* **7** (1987) 22-25.
3. (1) *Points to Consider in the Production and Testing of Interferon Intended for Investigational Use in Humans*, July 1980, revised 1 July 1982. (2) *Points to Consider in Monoclonal Antibodies for* In Vitro *Use*, 10 March 1982.

(3) *Points to Consider in the Manufacture and Testing of Monoclonal Antibody Products for Human Use*, June 1987. (4) *Points to Consider in the Production and Testing of New Drugs and Biologicals Produced by Recombinant DNA products*, 18 November 1983, revised 10 April 1985. (5) *Points to Consider in Cell Lines Used to Produce Biological Products*, 1 June 1984. Food and Drug Administration.

4. Guerra, J. and Finkelson, M. J. Validation of analytical methods by FDA laboratories. *Pharm. Technol. Mar.* **10** (1986) 74-84.
5. *Notes to Applicants for Marketing Authorization on Requirements for the Production and Quality Control of Medicinal Products Derived by Recombinant DNA Technology*. OECD Committee for Proprietary Medicinal Products, Draft 5, June 1986.
6. *Considerations for the Standardisation and Control of the New Generation of Biological Products*. The National Institute for Biological Standards and Control, London.
7. Giorgio, R. J. and Wu, J. J. Design of large scale containment facilities for recombinant DNA fermentations. *TIBTECH* **4** (1986) 60-65.
8. *Guideline on Sterile Drug Products Produced by Aseptic Processing*, Division of Drug Quality Compliance, National Center for Drugs and Biologics, Rockville, January 1985.
9. Waife, S. O. and Lasagna, L. From DNA to NDA—the impact of recombinant DNA technology on new drug development. *Reg. Toxicol. Pharmacol.* **5** (1985) 212-224.
10. Fry, D. M. FDA update on aseptic processing guidelines. *J. Parenteral Sci. Technol.* **41** (1987) 56-60.
11. Chapman, K. G. A suggested validation lexicon. *Pharm. Technol.* **7** (1983) 51-57.
12. Chapman, K. G. The par approach to process validation. *Pharm. Technol.* **8** (1984) 22-36.
13. PMA's Validation Advisory Committee. Process validation concepts for drug products. *Pharm. Technol.* **9** (1985) 78-82.
14. *Guideline on General Principles of Process Validation*. National Center for Drugs and Biologics and National Center for Devices and Radiological Health, Rockville, 3/29/83 (Draft 1); 3/7/84 (Draft 2); 9/85 (Draft 3); 3/86 (Draft 4).
15. *Coordinated Framework for Regulation of Biotechnology*, US Office of Science and Technology Policy, Parts I and II.
16. *Review and Analysis of International Biotechnology Regulations*. Arthur D. Little, Cambridge, MA, May 1986.
17. Duncan, M. E., Charlesworth, F. A. and Griffin, J. P. Regulatory requirements for licensing medicinal products of biotechnology. *TIBTECH* **5** (1987) 325-331.
18. Chew, N. Product development and registration, Part I: understanding FDA regulations. *BioPharm.* **1** (1988) 18-20.
19. Van Brunt, J. The valid way to quality biotech products. *Bio/technol.* **6** (1988) 120-122.

Appendix D *Column Packing*

Even when there is no extra-column zone spreading and columns with good flow distribution systems are used, there is always the possibility that the packing structure is worse in a larger column due to poor column packing methods or lack of information about the packing material. For example, modern rigid and semi-rigid packing materials that allow for high flow rates must also be packed at high flow rates to achieve optimal performance. These packing materials may be somewhat resilient and it is, therefore, necessary to lower

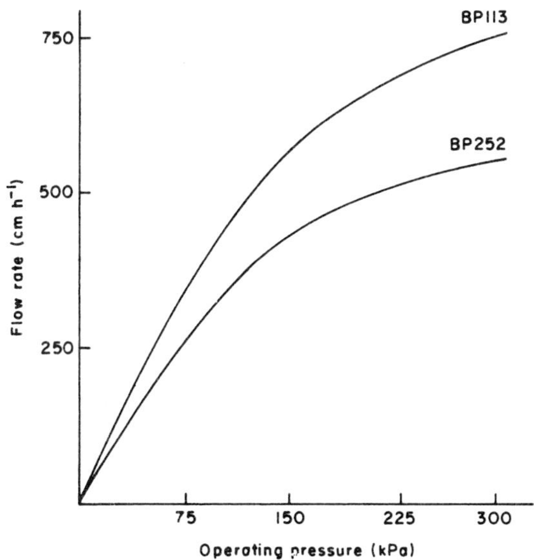

Figure 44. Pressure–flow rate curves for Q Sepharose® Fast Flow. BP113 Column; cross section, 100 cm^2; inlet, ¼″; bed height, 15 cm. BP252 Column; cross section 500 cm^2; inlet, ⅜″; bed height, 15 cm.

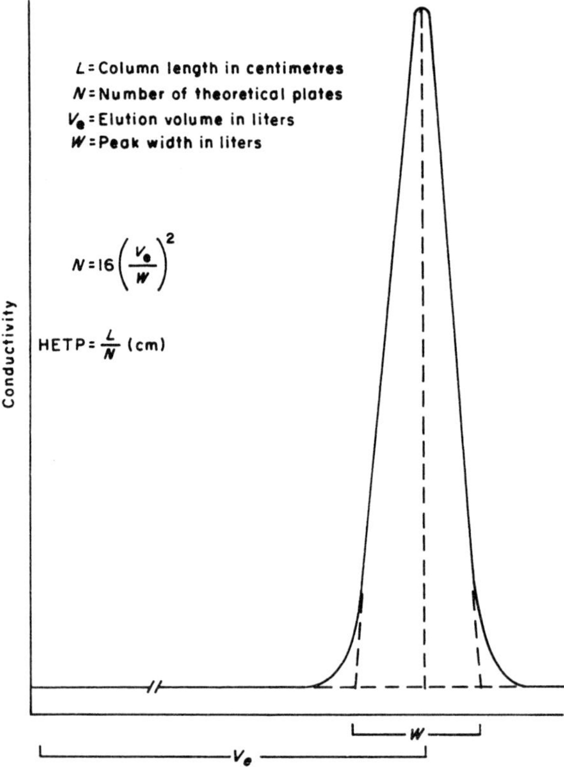

Figure 45. Column efficiency is best expressed in terms of height equivalent to theoretical plate (HETP).

the end piece or adaptor onto the gel surface as rapidly as possible after packing is completed. This will insure a tightly packed bed that will allow for good chromatographic performance.

Directions for packing each separation medium should be available from the manufacturer. In the event that packing directions are insufficient, a pressure/flow curve can be used to determine the best packing and maximum running flow rates. This is shown in *Figure 44*. After obtaining a pressure/flow curve, the separation medium should be unpacked and then repacked at a flow rate that is approximately 80% of the maximum flow rate. It is advisable to run the column at no more than 80% of the packing flow rate.

The user should be able to determine the status of the packing. Usually narrow zone samples of small molecules are used for this purpose. Typically, either sodium chloride or acetone is used. Sodium chloride is monitored using a conductivity meter, whereas a UV monitor is used to detect acetone. The conductivity is not linearly correlated with concentration, so one should be careful interpreting the results with sodium chloride or other salts. Acetone, on the other hand, interacts with some polymeric packings in such a way that the peaks are broadened by other factors than extra-particular axial dispersion. A sample containing concentrated (e.g. 10 ×) equilibration buffer may be used to test the packing of an ion exchanger.

The peak emerging in a test run of this kind is a measure of the increase in variance, and this test should be performed prior to use and on a routine basis. The elution profile is used to measure the efficiency of the column, which is dependent on how well the column is packed. HETP (height equivalent to a theoretical plate) is the best method of expressing the efficiency of a column (see *Figure 45*).

Index